Neighborhood as Refuge

26 in 26
Neighborhood Resource Centers
26 Neighborhood Strategies in a 26 month time frame
A Grant Funded by the LSTA
(Library Services & Technology Act)

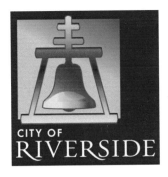

CITY OF
RIVERSIDE

Riverside Public Library

Urban and Industrial Environments
Series editor: Robert Gottlieb, Henry R. Luce Professor of Urban and Environmental Policy, Occidental College

For a complete list of books published in this series, please see the back of the book.

Neighborhood as Refuge

Community Reconstruction, Place Remaking, and Environmental Justice in the City

Isabelle Anguelovski

The MIT Press
Cambridge, Massachusetts
London, England

MIT Press books may be purchased at special quantity discounts for business or sales promotional use. For information, please email special_sales@mitpress.mit.edu.

This book was set in Sabon by the MIT Press. Printed and bound in the United States of America.

Library of Congress Cataloging-in-Publication Data

Anguelovski, Isabelle.
Neighborhood as refuge : community reconstruction, place remaking, and environmental justice in the city / Isabelle Anguelovski.
    pages   cm. — (Urban and industrial environments)
Includes bibliographical references and index.
ISBN 978-0-262-02692-5 (hardcover : alk. paper) — ISBN 978-0-262-52569-5 (pbk. : alk. paper)
1. Community development, Urban—Environmental aspects. 2. Environmental justice. 3. Urban ecology (Sociology) 4. Human ecology. I. Title.
HN49.C6A555   2014
307.76—dc23
                                                         2013028483

10 9 8 7 6 5 4 3 2 1

. . . (la forme d'une ville
Change plus vite, hélas! que le coeur d'un mortel)
(The form of a city changes faster, alas, than the human heart.)

—Charles Baudelaire, "Le Cygne," *Les Fleurs du Mal* (1857)

Contents

Acknowledgments

First, I am thankful to the neighborhood residents, community leaders, nonprofit organizations, planners, and policymakers in Barcelona, Boston, and Havana who welcomed me into their cities and communities at different stages of my research. They spent hours sharing their fascinating stories of work, hardships, and successes and taught me about loss, trauma, struggles, and accomplishments. I particularly thank José Barros, Jess Liborio, Penn Loh, and Travis Watson in Boston; Silvana and the volunteers of the Hortet del Forat in Barcelona; and the coordinators of the Casa del Niño y de la Niña and the Talleres de Transformación Integral del Barrio (TTIB) (Workshops for the Comprehensive Transformation of the Neighborhood) in Havana for their time, patience, and support. I am also grateful to the Department of Urban Studies and Planning at the Massachusetts Institute of Technology, the MIT-Spain Program, the Harold Horowitz Award, the Martin Sustainability Fellowship, the Center for International Studies, and the European Commission Marie Curie program, whose financial support has allowed me to pursue my research.

This project was initiated during my doctoral research, and it would not have been completed without the constant mentoring and reassurance of faculty members who supported me during my years at the MIT Department of Urban Studies and Planning. It is hard to put in words my gratitude for the warm, extensive, and yet demanding support from my PhD adviser, JoAnn Carmin. Her expertise in environmental justice and social movements has been immensely helpful, and the hours she has dedicated to my doctoral education and all-around mentorship cannot be counted. Her mentoring and guidance will remain a model of excellence for me for many years to come.

I am also particularly thankful to Lawrence Susskind, who originally encouraged me to conduct research on environmental rights and claims among marginalized communities. His teaching, practice, and research have motivated me and guided me throughout my professional and research development. As I built my research, Professor Susskind helped me to remain true to my interests and approaches while critically considering broader policy processes and changes and their implications for my work.

I also thank Diane Davis, who encouraged me to develop this study and offered many excellent suggestions for synthesizing, reframing, and sharpening my questions, methods, and argument. Her comparative work in urban sociology and international development was an early inspiration and will continue to inspire me in future projects. At MIT, I am also grateful for the support and feedback that I received from numerous professors, including Eran Ben-Joseph, Amy Glasmeier, Judith Layzer, Karen Polenske, J. Philip Thompson, and Lawrence Vale, and staff members, especially Kirsten Greco, and Janine Marchese, Sandy Wellford, and Karen Yegian.

During the course of my research, I had the opportunity to work with David Pellow from the University of Minnesota. His writings on environmental justice and environmental inequality formation were the intellectual grounding for my research and provided me with constant inspiration. He has helped me frame my work conceptually and theoretically in ways that I hope will advance the field of environmental justice. I also feel lucky to have received the mentorship of Joan Martínez Alier from the Universitat Autònoma de Barcelona for my postdoctoral work. Through many exchanges and writings about his work on the environmentalism of the poor, we worked together on the relevance of this theory in an urban context, which helped me refine my analysis of the role played by broader political economy processes in the creation of urban environmental inequalities. For their constructive feedback and encouragement, I also thank Gina Rey from the Instituto Superior Politécnico José Antonio Echeverría in Havana; Giorgos Kallis from the Universitat Autònoma de Barcelona; James Jennings and Penn Loh from Tufts University; Merilee Grindle, Archon Fung, Louise Ryan, and Rebecca Betensky from Harvard University; and John Forester from Cornell University.

I thank the wonderful friends and colleagues I met in Boston who were present for me during the development of this study. I also thank my longtime friends and colleagues in France and Spain—a special mention here to Céline Fercot. My family has been my most constant cheerleader and has provided the deepest support. I am grateful to my mother, father, and grandmother for instilling in me a belief in social change, political activism, justice, and dedication; my American family, Molly and Allen, for always listening and providing much patience and feedback throughout these years; and my German-Bulgarian family in Europe. My husband, Chris, deserves special thanks for his invaluable understanding and capacity to be a sounding board for my ideas and doubts, and for his sarcastic yet very much on point sense of humor. Finally, my daughter, Deia, brought daily smiles and laughter throughout this long writing process. I hope that the dedication and engagement of the activists presented here will one day inspire her to create a more environmentally just society.

1

Introduction

Two Vignettes on Loss, Trauma, and Survival

Experiences of Refugees in an Urban Greenhouse
Every Friday, a group of East African refugees walks energetically into the grounds of The Food Project greenhouse on Brook Avenue in the Dudley section of Roxbury, one of the twenty-one official neighborhoods of Boston, Massachusetts. The greenhouse is a 10,000-square-foot space that acts as a community space and learning center for residents and gardeners. Appearing shy and cautious at first, refugees soon engage in lively conversations about their raised beds, the workshop on safe farming they attended the previous week, and the growth of their greens and herbs in winter. These refugees are one of nine groups that have spaces in the twenty-seven community bays available at the greenhouse. The gardeners are not alone in the greenhouse but are supported by attentive Food Project staffers and clinicians from the Boston Center for Refugee Health and Human Rights, who are helping them to heal from traumas experienced in their home countries. Beyond the well-known benefits of urban agriculture for increasing access to healthy, fresh, and affordable food and for building stronger ties between people, the greenhouse project helps refugees deal with traumatic experiences, learn about their new city and neighborhood, and build a new home for themselves and their families in Dudley. It is a beautiful green bubble of tranquility and safety for the community and a nurturing space for residents.

While the participants plant seedlings, receive technical advice, and work in the greenhouse, they learn about themselves and each other. Every week, they walk with their clinician down Brook Avenue from the nearby Boston Medical Center and discover their neighborhood and its history. At the beginning of the project, many women felt unsafe in the neighborhood and uncomfortable walking alone. As they have learned about new landmarks on their weekly walk, explored corners of the neighborhood, met friendly faces, and finally arrived at the greenhouse to work on their crops, their sense of safety has grown. The greenhouse also helped them strengthen their relationship with their clinician. In this space, they have developed a greater sense of security and nurturing and do not feel intimated or rushed, as they might be by the more impersonal atmosphere of a hospital. A feeling of greater trust and a sense of proximity have developed between the farmers and their clinician. The greenhouse project has thus played a therapeutic role for the refugees by helping them slowly feel at home in their new city and overcome traumatic experiences of war or crisis. It is a safe haven for them in multiple ways.

Because many gardeners have chosen to grow vegetables from their own country, they have also reconnected with their home cultures and traditional practices and shared them with other refugees and participants. Rather than feeling like outsiders and foreigners in a new country, they are learning how to live, survive, and thrive in a new city through the medium of food by growing food and by teaching others about food. In addition, growers share the space with other groups that have beds in the greenhouse, including senior resident groups, nonprofit organizations, a mosque, and local public schools, and these groups learn from one another. Some refugees have led workshops about Somalian cooking, and others have organized a tea celebration based on a Ugandan-type of lemongrass. By talking about their cultural roots, participants also recover from the loss of leaving their homeland or home region and build a new and stable life for themselves. These workshops are part of The Food Project's Grow Well, Eat Well, Be Well, Cook with Your Neighbor initiative. Its "Be Well" aspect embodies the vision of a greenhouse as a space for community members to come together, build closer bonds, and learn about each other through the medium of food production and preparation.

Experiences of Youth in an Urban Farm
The greenhouse is not the only project that The Food Project has developed to provide residents with a sense of nurturing and help them recover from traumatic life experiences and losses. The Academic Year Program (AYP), a follow-up to The Food Project's successful Summer Youth Program, welcomes local youth on Saturdays and after-school hours to a variety of training and mentoring activities. Youth participants farm in three lots in Dudley, help at biweekly farmers' market at the intersection of Dudley Street and Blue Hill Avenue, organize volunteers in the farms, and speak at different events and fundraisers. Participants learn about

food systems and farming and also develop leadership and public speaking skills. Most important, as Alexandra, the AYP coordinator, explains, the curriculum helps these young people to process trauma. She notes that many youth of color in Boston suffer from trauma—either at an individual level (through family tensions or uprooting) or at the collective level (through the degradation and abandonment of their neighborhoods and the territorial stigma attached to them). As young people develop and try to build a better future for themselves, they must overcome these negative feelings of placelessness and loss.

Historically, Dudley has welcomed large waves of African American migrants and immigrants, most recently from Cape Verde, Central America, and West and East Africa. However, the Dudley section of Boston's Roxbury neighborhood has been ravaged by decades of arson, illegal trash dumping, and abandonment by public authorities and investors. In the 1980s, Dudley had 1,300 empty lots and looked like a no-man's land devastated by an urban war. Fifty-four sites were considered hazardous, contributing to high rates of lead poisoning in children and other kinds of contamination (Shutkin 2000; Settles 1994). The neighborhood also lacked green space, playgrounds, and other recreational facilities. From a nutritional standpoint, the abandonment of the neighborhood by supermarkets and other grocery stores had converted it into a food desert. However, by the end of the 1980s, residents and organizations had begun to address these substandard environmental and health conditions. After organizing the Don't Dump on Us campaign and winning three lawsuits against businesses that were illegally contaminating the land, they worked to enhance the environmental quality and livability of their neighborhood.

The Food Project has been a major actor and activist throughout the reconstruction of this neighborhood. Its staff members have helped residents grow healthy and affordable food by producing 250,000 pounds of chemical-pesticide-free food. Most important, however, they have worked with youth. The leadership programs organized by The Food Project mentor children and adolescents in overcoming feelings of loss and grief, making conscious choices about their futures, and protecting the assets of their neighborhood—especially its land. In weekly activities, young participants learn to put negative experiences aside and focus on building new opportunities for themselves and their neighborhood. Being in a warm and caring environment in the urban farms or farmers' markets, they can process trauma and placelessness.

In other words, The Food Project's greenhouse and farming programs for youth help residents to collectively and individually heal their wounds and rebuild their community for themselves and their families. Long-term residents overcome the traumatic experiences of having lived for many years in a decaying neighborhood, and newer residents can heal feelings of loss after having been uprooted from their homeland or community and resettled into a new (and fragile) community.

Overcoming Long-term Environmental Degradation and Abandonment

Life in historically distressed neighborhoods is often closely coupled with degraded infrastructure, substandard services, unhealthy housing structures, and severe environmental hazards. These low-income neighborhoods generally receive fewer environmental amenities (such as pleasant open spaces) and services (such as street cleaning) than wealthier communities receive. Those wealthier (and often whiter) communities tend to benefit from environmental privileges in the form of parks, coasts, and forests, often in a racially exclusive way (Landry and Chakraborty 2009; Park and Pellow 2011). To outside eyes, distressed neighborhoods often appear bleak and abandoned with much unused vacant space. Many storefronts are closed, sidewalks are deserted after dark, and open space is unused due to an experienced or a perceived lack of safety. In the United States, decades of government and private disinvestment together with urban-renewal policies have contributed to the decline of inner-city neighborhoods. In Europe, lower-income residents and immigrants often settle in older inner-city neighborhoods and low-income suburbs, where they have often been left behind by authorities. In some cases, their neighborhoods are redeveloped in ways that privileged demolition and rebuilding rather than revival. In the global south, dilapidated urban neighborhoods have often benefited from public attention and investment when they offer historic and touristic cultural options. Often, however, their residents have had to relocate.

Today activists within historically marginalized urban communities are organizing against long-term abandonment and neighborhood degradation. In their initiatives, residents, community groups, and nonprofit organizations prioritize accessible green and recreational spaces, urban gardens and farmers' markets, walkable communities, green and healthy housing, and improved waste management. This mobilization takes places in a variety of cities around the world, independent of levels of democratization, development, and urbanization. Examples include the growth of urban farms and community gardens in Detroit and Los Angeles where there once were foreclosed abandoned houses and vacant dirty lots; the creation and enhancement of green and recreational spaces in the shanty-town of Villa Maria del Triunfo in Lima, Peru; and the community initiatives for improved waste collection and composting in rapidly growing cities like Mumbai, India. Despite the fragile socioeconomic conditions of some communities and families, residents actively participate in the revitalization of their neighborhood and have received widespread support from local environmental nongovernmental organizations (NGOs), community-based organizations, small neighborhood groups, and public and private funders.

Community organizing is not a short-term activity, and it requires a commitment to neighborhood engagement. Local activists fight "brown" contamination and unwanted waste sites, but also work to improve the overall environmental and health conditions of a neighborhood over the long term. Ongoing community-based and community-supported initiatives are vital for improving and sustaining environmental equity at the local urban level. In other words, in many instances residents are committed to improving long-term livability and unite to protect the environmental quality of their neighborhood beyond protests against specific threats, risks, and pollution sources.

Projects such as urban farms, community parks, fresh food markets, bike paths, green public transit, and playgrounds and sports grounds fulfill multiple roles at once. They are vehicles for improving the livability of urban neighborhoods, creating healthy communities, decreasing criminality, enhancing safety, and strengthening local urban planning and democracy practices (Bell, Wilson, and Liu 2008; Diaz 2005; Gottlieb 2005; Takano and Tokeshi 2007; Birch and Wachter 2008; Kuo and Sullivan 2001; Agyeman and Evans 2003; Corburn 2009; Shutkin 2000; Bullard 2007). Today it would be hard to find critics or opponents of initiatives that pursue these different components of urban just sustainability (Agyeman, Bullard, and Evans 2003; Agyeman and Evans 2003). However, in many cases, neighborhood transformation has taken many years, much advocacy and on-the-ground work, and the participation of a broad range of supporters. Sometimes neighborhood advocates have suffered setbacks. Today, "green gentrification" (the perceived or lived displacement of traditional residents from the neighborhood as it becomes more livable) is real in some cities (Checker 2011; Curran and Hamilton 2012; Gould and Lewis 2012; Pearsall 2012).

Community-based engagement in the global north and south suggests that caring for and enhancing the long-term environmental quality of one's home is not a function of wealth, political systems, or contexts of urbanization. Nor is it a function of imitating trends or following funding opportunities (community action often started when green movements for sustainability in the inner city were just beginning). Residents are not following "greening" trends and agendas. Their battles also challenge assumptions that residents of degraded neighborhoods are eager to move to wealthier areas with better services, housing, and environmental conditions. They question policies such as the deconcentration or dispersal of poverty (Goetz 2003; McClure 2008; Turner 1998; Aalbers, van Gent, and Pinkster 2011; Kearns 2002), which provides incentives for residents to move to neighborhoods with allegedly greater opportunities and creates conditions for greater socioeconomic diversity.

Why do similar local patterns of concern, mobilization, and achievement arise in cities with different political systems and contexts of urbanization? What motivates residents to take action in their neighborhood and remain committed to it over time despite the dire baseline conditions of the place and the many obstacles to transforming it? Such questions motivated this study. After working or conducting research in different cities in North America, Latin America, and Europe, I found myself both surprised and puzzled by comparable patterns of neighborhood-based activism.

Urban Environmental Justice at a Crossroads: Bringing in Place and Community

Thanks to the long-time commitment of urban planning scholars, sociologists, epidemiologists, and geographers to the study of environmental inequalities in American cities, the disproportionate exposure of urban low-income groups and communities of color to environmental toxins and contamination is well known. Environmental injustices are widespread in urban distressed communities, and despite executive orders, local laws, municipal ordinances, and numerous local victories, they do not cease to exist. Such communities are also less likely to receive environmental services and amenities, have access to healthy food markets, and engage in sports and other physical activities. In Europe, studies of environmental inequality are growing, and as environmental justice (EJ) scholarship in the United States has accomplished, they have documented the proximity of lower-income communities to refineries and chemical facilities and to substandard environmental and health conditions more generally. In the global south, environmental injustices manifest themselves in more subversive ways as poor or native communities suffer from environmental contamination and resource extraction in fragile ecosystems. At best, there is very little state control over polluting and extractive activities, and at worst, there is total permissiveness. In addition, the political economy of extraction, production, and waste translates into the export of electronic and material waste from northern corporations, states, and cities to southern cities and ports. Furthermore, cities are divided into into wealthy, often gated and lush neighborhoods and shanty neighborhoods and unauthorized slums populated by rural migrants and low-income families who are not connected to city services such as waste collection or water provision.

Low-income and minority communities have not remained passive in the midst of these growing environmental and health issues. Local activists have raised their voices against inequalities and injustices. They struggle for civil and human rights, social and economic justice, and a fair distribution of ecological resources. The common denominator of community demands is the definition of *environment* as a place where people live, work, learn, and play together rather than as a wild ecosystem that needs to be protected by all means without considering people's basic needs and relationship to this environment. Recent surveys conducted in low-income and minority neighborhoods show that residents consider violence, healthy food, and affordable housing as more salient health issues than pollution (Corburn 2009). More recently, urban residents have associated demands for improved livability and environmental quality with a right to the city and, more generally, spatial-justice demands.

People indeed develop tight connections to their place of residence or neighborhood. Neighborhoods not are neutral repositories of their struggles and they are places that are imbued with meanings and associations connected to tradition, identification, and experiences. Residents of urban distressed communities develop attachments to their neighborhood because of the networks, relations, and affective bonds that they have created and that have often helped them confront stressful situations and build resilience. Despite official images and reports about the plagues of inner-city neighborhoods, residents create an active village life around public spaces and cherish the cultural and social activities they organize.

However, distressed neighborhoods are often at risk of disruption and fragmentation by degradation and decay, urban renewal projects, and gentrification. To confront such changes, residents construct autonomous images of place and community (such as cultural and social events) that contest the stigmas of esthetic degradation and low environmental quality. They also take actions to protect their place and assets. For instance, residents invest time in the neighborhood, attend public meetings about municipal projects for the neighborhood, and participate in protests to oppose new developments.

Yet, our understanding of how place, place attachment, and sense of community shape struggles for long-term environmental quality and livability in urban deprived neighborhoods is rather limited. Little is known about the role played by place and identity in the organization of local activists who fight for environmental improvements in marginalized neighborhoods and about the ways that concerns for place connect to environmental and health initiatives. What forms of connections do activists develop with marginalized neighborhoods that have suffered from substandard environmental conditions? How do their experiences in this neighborhood shape activists' engagement in environmental revitalization projects? How do parks, playgrounds, community gardens, and healthy housing connect to underlying goals? Do they use their accomplishments as tools to advance other claims related to their place, its stability, or its transformation?

Until recently, environmental justice scholarship has focused on environmental "bads." However, more and more attention is now given, especially in the United States, to community-led or -advocated green projects. In this book, I examine the physical transformation of neighborhoods through environmental and health projects and the ways that such changes connect back to the remaking of place for minority and low-income residents. How do local activists grapple with issues of neighborhood attachment, place reconstruction, and sense of community? I hope to identify how activists use environmental and health improvements to construct a different relationship to their place and a different relationship to the political institutions, processes, and actors that shape it.

A Comparative Study of Neighborhood-Based Environmental Revitalization

Study Design

This book is based on a comparative study of three neighborhoods that have a large proportion of low-income residents of color—the Casc Antic neighborhood in Barcelona, Spain; the Dudley neighborhood in Boston, Massachusetts, United States; and the Cayo Hueso neighborhood in Havana, Cuba. The focus of this study is on parks and playgrounds, sports courts and centers, urban farms, farmers' markets and healthy food providers, waste management, and healthy housing projects.

While doing preliminary fieldwork in each city and assembling academic articles, reports, and newspaper resources about them, I found that the Casc Antic, Dudley, and Cayo Hueso were emblematic cases of community organization, action, and achievement in regard to improvements in environmental conditions over the past two to three decades. These neighborhoods are recognized in their cities for their local activism on issues connected to livability and environmental quality. Their engagement has brought about environmental revitalization, and my initial fieldwork and my inductive approach to research revealed common patterns and experiences of neighborhood-sponsored environmental revitalization in cities that at first glance do not share many attributes (Barcelona, Boston, and Havana). I was struck by similarities in discourses as I met residents and leaders in each city. After considering other neighborhoods (including Atarés and Pogolotti in Havana; Dorchester, East Boston, and Mattapan in Boston; and the Barceloneta, Raval, and Nou Barris in Barcelona), discussions with community activists and planners led me to conclude that these other places were not representative of the long-lasting changes, comprehensive neighborhood improvements, and dynamics of community engagement around environmental revitalization that I encountered in the Casc Antic, Dudley, and Cayo Hueso. These three neighborhoods are situated in the center of their respective cities, so they are comparable to each other in terms of geographic location within the city, general infrastructure, historical relevance, and physical proximity to decision makers, planners, and economic players.

Even though Havana is part of a socialist, centralized, and autocratic regime while the two other cities are part of democratic systems, I included Havana following an inductive and dynamic approach to research. The similarities in discourses and tactics of residents and leaders in Barcelona, Boston, and Havana led me to believe that a comparative analysis of their mobilization over time would be powerful because residents showed similar commitment to their neighborhood in a variety of political systems and contexts of urbanization. I was thus committed to maximizing the diversity of political systems, contexts of urbanization, and histories of marginalization to examine in greater detail how these different conditions affect the claims, struggles, and strategies of historically distressed neighborhoods.

My goals in this book are partly theoretical and conceptual and partly methodological and epistemological. I aim to further existing theories of environmental justice while contributing to nascent research in comparative urbanism beyond an opposition of global north and south.

I attempt to provide a refined analysis of cities with comparative spatialities, connections, and processes, (McFarlane 2010; Robinson 2011) and examine the growth of community-based organization for greater livability and environmental quality in degraded and fragile urban neighborhoods around the world. The large majority of environmental justice studies are based on one site or city or on several cities within one country. Some edited volumes focus on large geopolitical areas (Agyeman and Ogneva-Himmelberger 2009; Carruthers 2008). While such studies have provided important analysis about the distribution of and battles against environmental inequalities and unjust planning decisions, they leave environmental justice research compartmentalized and divided by regions or continents.

What drives activists in urban marginalized neighborhoods around the world to work to improve environmental quality and livability? How do questions of place and people's sense of community play out in local environmental justice struggles? Traditional social science research tends not to consider some of these issues—how neighborhoods and communities in dissimilar economic, cultural, and political settings mobilize against similar environmental and health challenges and situations of marginalization; what commonalities researchers might find across cases to explain similar positive outcomes, processes, and changes; and how variations and differences manifest themselves. After my preliminary fieldwork, I decided to engage in an in-depth research project to examine how residents and their supporters organize to work toward urban environmental revitalization.

The scholarly field of environmental justice was first developed in the United States. European research, for example, has barely begun to analyze community organization around urban livability in marginalized neighborhoods, despite the prevalence of the issue in local politics. Studies center mostly on identifying, through quantitative analyses, the disproportionate environmental harms that are experienced by immigrant and poor populations, as in traditional EJ studies in the United States. While researchers in the global south have undertaken numerous environmental justice studies, they generally focus on rural ecosystems, contamination, and resource extraction. Although the categories and framework used by traditional environmental justice scholars do not match perfectly with the European context and with southern societies, the weakness of comparative empirical research makes it difficult to elaborate original models that reflect the circumstances of other societies and allow for rigorous comparative research across contexts.

Last, by examining how responses to environmental degradation help build participatory spaces in neighborhoods, this book also contributes to policy-relevant knowledge of the types of community organization that allow vulnerable communities to influence planning decisions in a variety of political and socioeconomic contexts. I look at how activists within degraded neighborhoods in dissimilar economic, cultural, and political settings mobilize against similar environmental and health challenges; receive support for their community projects; and leverage power and at times alliances in their relationships with decision makers.

There are various ways to define political systems and regimes. For the purpose of this study, I classify systems according to level of guaranteed political rights, citizen participation, protection of civil liberties, and democratic elections (Tilly 2006). According to these criteria, Boston represents a well-established democracy with historic roots of civic engagement, high protection of liberties, and regular elections at multiple levels. Barcelona represents a young democracy that was reestablished in 1978 after half a century of dictatorship and has a growing level of civil liberties and citizen participation. Havana is an example of an autocratic and centralized government that has weak opportunities for citizens' engagement in decision making, low respect for civil liberties, and one-party elections at the national level.

The contexts of urbanization vary greatly among these cities. Boston is a wealthy, well-developed, and established city with twenty-one strong neighborhoods that make up its social and economic fabric. It also has experienced racial violence, segregation, and marginalization of poor, ethnic, and immigrant areas. Barcelona, a dynamic and wealthy city, received little investment and attention under Francisco Franco's dictatorship (1936–1975) but has seen large development projects, new infrastructure, and strong protest movements since the return of democracy. Many areas of the city have been redeveloped and transformed. Over the past few decades, the city has also welcomed waves of migrants from Andalucia and Galicia and immigrants from Latin America and North Africa who have integrated into the fabric of the city despite initial clashes and tensions with older residents. Finally, Havana is a formerly prosperous capital city in a developing country. After the communist revolution led by Fidel Castro in 1959, the country tried to eradicate racism and discrimination against Afro- Cubans and provide them with education, health care, and housing. For thirty years, the Soviet Union was Cuba's primary source of investment, and when the USSR fell in 1989, Cuba suffered a deep economic and social crisis, from which it is still recovering. In Havana,

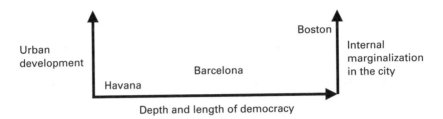

Figure 1.1

Case site selection for Boston, Barcelona, and Havana: Democracy, development, and marginalization

centrally located neighborhoods such as Cayo Hueso have decayed over the last forty years, and the authorities have been mostly unable to respond effectively to urban neglect, except in the neighborhood of Habana Vieja. Figure 1.1 summarizes the selection of the three case studies and situates the three cities in which they are located along three axes: levels of urban development (from low to high), depth and length of democracy (from low to high and short to long), and internal marginalization (from low to high) in the city.

An Overview of Dudley (Boston), Casc Antic (Barcelona), and Cayo Hueso (Havana)

Dudley, Casc Antic, and Cayo Hueso are three centrally located historically marginalized neighborhoods in Boston, Barcelona, and Havana. Their baseline conditions and transformation are summarized in table 1.1.

In Boston, the Dudley neighborhood is situated within the broader official neighborhood of Roxbury, less than two miles from downtown (figures 1.2 and 1.3). The majority of its residents are lower-income African Americans, Cape Verdeans, and Latinos. It is named after the street that runs from Dudley Station on Washington Street to Uphams Corner in North Dorchester, but the neighborhood is larger than the well-known Dudley Street triangle. In the 1980s, it was a violence-ridden, arson-devastated area of the city. By 1984, 21 percent of Dudley (1,300 parcels) was vacant land and waste-filled properties, being the victim of municipal abandonment and arson by mostly white property owners who were eager to leave the inner city (Layzer 2006). In 1990, one out of two Dudley children lived below the federal poverty level of $12,700 for a family of four. Between 1987 and 1991, the neighborhood had a disproportionately high level of homicide deaths: thirty-one people were killed by homicide

Table 1.1
Dudley, Casc Antic, and Cayo Hueso: An overview

Dudley (Boston)	Majority of low-income African American, Cape Verdean, and Latino residents
	About 1,300 vacant lots by the mid-1980s, the majority of them contaminated and abandoned by the city of Boston and affluent property owners
	Illegal trash-transfer stations
	Lack of parks and recreational facilities
	Food desert
	50% of children below the poverty line
	Since 1984, community-led land cleanup, management, and redevelopment of parcels into urban farms, gardens, community gyms, and healthy food businesses
Casc Antic (Barcelona)	Residents 31% foreigners and a majority in poverty
	Legacy of Franco: Crumbling housing, poor waste management, and abandoned and unsafe public spaces
	The 1980s: PERIS urban plans, unequal development, negative social and environmental effects
	Urban conflicts since the end of 1990s (such as Forat de la Vergonya), occupation and reconstruction of abandoned park
	The 2000s: Community-based environmental revitalization projects and advocacy for improved socio-environmental conditions
Cayo Hueso (Havana)	Dense and predominantly Afro-Cuban neighborhood in Centro Habana
	By 1989: Degradation of buildings and sanitation and further decay during the Special Period crisis
	More than 50% of residents without daily access to potable water
	Few green areas and safe public spaces
	The 1990s and 2000s: Workshops for the Comprehensive Transformation of the Neighborhood promoted by the Grupo para el Desarrollo Integral de la Capital planning agency as autonomous community-based revitalization projects
	The 1990s-2000s: Independent resident projects around public space and recreational and sports facilities

Figure 1.2

Aerial view of the Dudley neighborhood's central area in Boston.
Source: Google Maps

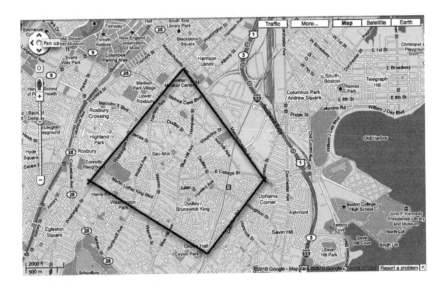

Figure 1.3

Boundaries of the Dudley neighborhood in Boston.
Source: Google Maps

in the Dudley core area (population 12,000), and 481 were killed in the entire city of Boston (population 574,000) (Medoff and Sklar 1994). The abandonment of the neighborhood by supermarkets and other grocery stores had also transformed Dudley into a food desert.

Starting at the end of the 1980s, residents, environmental NGOs, and community organizations used environmental community policing strategies, innovative institutional arrangements with the city, and multiscale coalitions to transform hundreds of empty lots into urban farms, neighborhood parks, fresh food markets and bakeries, and sports centers and playgrounds. First, they organized against illegal dumping through the Don't Dump on Us campaign. After the city of Boston granted the power of eminent domain to the Dudley Street Neighborhood Initiative (DSNI), the group gained control over 1,300 parcels of abandoned land. Working with Alternatives for Community and Environment (ACE), a local environmental NGO, DSNI closed two illegal trash-transfer stations and won three legal cases of illegal dumping. After these victories, DSNI, ACE, and The Food Project established lead poisoning prevention projects in 160 urban gardens. At a more decentralized level, residents founded neighborhood associations and created community gardens throughout Dudley, such as the Savin and Maywood Streets Community Gardens and the Virginia and Monadnock Streets Community Gardens. They also developed sustainability projects, including rain barrels, composting, and raised beds.

Since the 1990s, opportunities for healthy food and physical activity have greatly increased. In 1992, The Food Project created three urban farms and later developed biweekly farmers' market and a green house. Today, Dudley Square itself is home to the Haley House Bakery Café. In addition, community organizations have done much work on green and public spaces, especially through the reconstruction and creation of parks and seven tot lots and playgrounds. The nonprofit Boston Schoolyard Initiative created new schoolyards and outdoor classrooms at three schools, which are also accessible to all the residents living in the area. New community centers, multipurpose facilities (like the Kroc Center), and a community gym (Body by Brandy) have also emerged.

In Barcelona, Ciutat Vella, the district where Casc Antic is located, is a labyrinth of small streets and plazas, with only a few streets measuring more than seven meters in width (figures 1.4 and 1.5). Small stores and workshops usually occupy the ground floor of buildings, and the upper floors are used for housing and offices. Street names—such as Cotoners (cotton makers) and Flassaders (blanket makers)—reflect the traditional

Figure 1.4
Aerial view of the Casc Antic neighborhood in Barcelona.
Source: Veins en Defensa de la Barcelona Vella

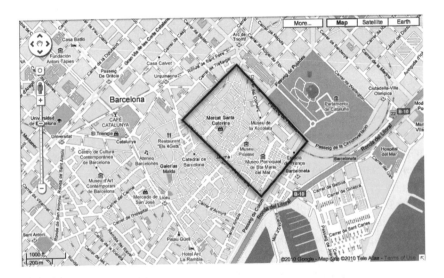

Figure 1.5
Boundaries of the Casc Antic neighborhood in Barcelona.
Source: Google Maps

artisans who established themselves in the neighborhood. Located between Barcelona's waterfront and the broader streets of the newer Eixample district, Casc Antic is a lively, diverse, and dense neighborhood. It has traditionally been a neighborhood that welcomed waves of national and international migrants. In the mid-2000s, 31 percent of the Casc Antic residents were foreigners, many of them illegal immigrants without regular income, and today many residents live just above or under the poverty line, particularly in the areas of Sant Pere and Santa Caterina.

In the early 1980s, after forty years of dictatorship, the neighborhood was suffering from crumbling housing units, poor waste management, and abandoned public spaces. While the newly democratic municipality invested resources into revitalizing degraded areas and promoting tourism, this process displaced 2,000 residents, destroyed 1,078 buildings, did not resolve the problems in waste collection and management, and did not create recreational, sports, and green areas (Ajuntament de Barcelona 2005). Residents still talk today about unequal revitalization and about feeling cheated. By 2001, only 58 percent of the remaining buildings were adequately maintained against 80 percent in Barcelona as a whole (Martín 2007).

By the middle of the 1990s, citizens were speaking out against excessive tourism, the conversion of the city into a theme park, real estate speculation, extended gentrification, and deficient social and urban infrastructure (Capel 2007; Borja, Muxí, and Cenicacelaya 2004; Taller contra la Violencia Inmobiliaria y Urbanística 2006). In 1997 and 1999, Promoción de Ciutat Vella SA (PROVICESA), the company that was in charge of the old town reconstruction, evicted residents and demolished old buildings with a plan to build a car parking lot and high-end apartments (Alió and Jori 2010). When the contractors abandoned the area, residents and their supporters (including squatters, architects, and neighbors' associations) rebuilt the empty spaces into community gardens, a green zone, a large plaza with playgrounds, and soccer and basketball fields. In 2007, after years of conflict, the municipality agreed to reconstruct the place that residents had baptized "Forat de la Vergonya" (Hole of Shame) into a permanent multiuse green space and to support the development of a new community garden—the Hortet del Forat. For 2.8 million euros, the reconstruction also included the renovation of water, gas, sewage, and public lighting infrastructure. Community organizations also pushed the municipality to create small green spaces and to renovate the Allada Vermell plaza to transform it into a greener and more child-friendly public space.

Several community-based organizations and projects emerged after years of confrontation in the neighborhood, including a network of community-supported agriculture (CSA) groups and an ethical trade and exchange market. The Fundació Ciutadania Multicultural started Mescladis, whose members teach cooking workshops and train immigrants in healthy food preparation and sales. In addition, residents and community groups improved waste management through educational campaigns in apartment buildings, and they advocated for the creation of a municipal minirecycling center and the construction of a pneumatic waste system. Coupled with the the construction of a large municipal sports center in 2010, leading sports leagues—such as A. E. Cervantes–Casc Antic (AECCA), Fundació Adsis, and Fundació Comtal—are training at-risk youths throughout the Casc Antic.

Cayo Hueso is one of Havana's oldest, most vibrant, and diverse neighborhoods (figures 1.6 and 1.7). It is a predominantly Afro-Cuban neighborhood in the center of the city in the municipality of Centro Habana, which has the highest density in the entire country: 170,000 inhabitants squeeze into 3.5 square kilometers. Cayo Hueso's approximately 40,000 residents live in 0.83 square kilometers (Spiegel, Bonet, Yassi, Molina, Concepcion, and Mast 2001). An overhead view of Centro Habana shows a maze of interconnected streets, passages, and courtyards facing the beautiful Bay of Havana. At ground level, on the streets and inside buildings, games, parties, music, festivities, and other activities give the impressions of microneighborhoods within the neighborhood. Its architecture is diverse and includes neogothic religious buildings, neoclassical palaces and offices, art deco apartments, and art nouveau houses (Gómez and Nieda 2005). In Cayo Hueso, streets are not merely passageways but function as places of daily encounters and activities. The distinction between private space and public space both inside and outside buildings is loose.

Until the late 1980s, most buildings in Cayo Hueso were highly degraded and abandoned by the authorities. Ninety-four percent of its buildings were considered in a "very bad" state (against 45 percent in Centro Havana as a whole), with limited access to potable water and damaged waste disposal systems (Spiegel et al. 2001). Residents had little access to green and public spaces and playgrounds (0.22 square meters per inhabitant against 3.8 square meters for Havana as a whole), and seventeen out of twenty-one recreational and green spaces were in bad condition. Schools did not have much space for physical activity. After the economic crisis that began in 1989, drug dealing, domestic violence, and prostitution became symbols of the neighborhood's degradation and neglect (Yassi, Fernandez, Fernandez, Bonet, Tate, and Spiegel 2003).

Figure 1.6

Aerial view of the Cayo Hueso neighborhood in Havana.
Source: La Ciudad Universitaria José Antonio Echeverria (CUJAE)

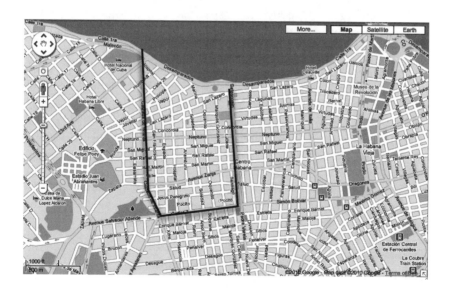

Figure 1.7

Boundaries of the Cayo Hueso neighborhood in Havana.
Source: Google Maps

Yet the hardships caused by the post-Soviet "special period" have inspired residents to participate in neighborhood revitalization initiatives through the Talleres de Transformación Integral del Barrio (TTIB) (Workshops for the Comprehensive Transformation of the Neighborhood). These workshops for neighborhood transformation are sponsored by the Grupo para el Desarrollo Integral de la Capital (GDIC) (Group for the Comprehensive Development of the Capital), a decentralized planning agency, but are led by community members. The TTIBs represent an alternative model of decentralized and participatory urban development that promotes the reappraisal of neighborhoods as clearly identified territories (Hearn 2008). In Cayo Hueso, the TTIB repaired twelve tenement buildings, created twenty family medical clinics, converted empty lots into parks and recreational areas, developed tree replanting and street beautification projects, and fixed up parks (such as Parque Trillo and Parque Maceo). The TTIB also persuaded the municipality to increase the number of trash collectors and containers and improve street cleanliness. Finally, TTIB members successfully fought for the creation of a high-yield urban farm.

Independent leaders and organizations also have started initiatives in Cayo Hueso that support green-space development, street beautification, and urban agriculture. In 1998, the United Nations Children's Fund (UNICEF) sponsored the construction of the Casa del Niño y de la Niña as a new community and recreational space that involves children in recreational activities, neighborhood environmental clean-up activities, and community health campaigns. Two independent leaders also created a new community gym (Quiero a Mi Barrio) and rehabilitated an outdoor sports ground (El Beisbolito) for martial arts and baseball. Recently, the nongovernmental organization Save the Children started renovating school playgrounds and sports facilities. In the urban agriculture area, residents initiated permaculture projects in housing complexes. Finally, a local artist named Salvador joined with neighbors to create the Callejón de Hamel—a street that they converted into a park with artwork, playgrounds, trees, and murals.

Methods

The data collection for this study is based on one hundred fifty semistructured interviews that I conducted in Barcelona, Boston, and Havana; dozens of unstructured interviews; and observations, participant observations, and secondary data collection. This fieldwork was conducted in two stages in each city in 2008 and 2009 and complemented by targeted interviews and document collection in 2012 to verify and update the data. Appendix D presents the data-analysis stages and the techniques that were used for this research (including grounded theory, process tracing, and analytic and historical narratives).

First, I interviewed community organizers and program managers from neighborhood organizations, representatives of local NGOs that were working to improve local environmental conditions, and active residents and leaders in each neighborhood. Second, I conducted a few interviews with residents, workers, and professionals (that is, political figures, social workers, representatives of business associations, and members of social or cultural associations and groups) who were not active in the neighborhood but whose perspective helped me understand some fundamental aspects of Dudley, Cayo Hueso, and Casc Antic—their stakeholders, groups, internal tensions, and relevant experiences. Third, I interviewed officials and professionals from municipal agencies and offices (including planning offices, neighborhood development departments, environmental departments, and public health departments) whose work, positions, and opinions influenced the organization and mobilization of marginalized urban neighborhoods. Finally, I met with NGOs and funders whose support to Dudley, Casc Antic, and Cayo Hueso appeared to have been decisive in the pursuit of environmental and health projects. I attempted as much as possible to acknowledge that several core identities and projects might coexist with, compete with, or support others within each neighborhood. Whenever possible, I also interviewed different types of leaders in each neighborhood—unofficial moral and religious leaders, officially elected leaders, leaders occupying important official positions, and members of local organizations. Sampling was done through snowball techniques. Appendixes A, B, and C list the people who agreed to participate in semistructured interviews with me in the three neighborhoods. For confidentiality purposes, some of the names have been changed.

I also observed meetings and events and engaged in participant observation of environmental and health projects by volunteering at one community organization in each of the three neighborhoods. These methods helped me to comprehend the goals and visions of residents and workers, the development of their neighborhood initiatives, and the hurdles that residents encountered while trying to achieve their goals. I was interested in understanding the spirit and context in which projects took place and the dynamics between participants. Volunteering helped me feel more at ease with each neighborhood and better reflect on the changes that it had experienced or was experiencing. In Boston, I volunteered at the urban farms of The Food Project, at an environmental justice organization called Alternatives for Community and Environment (ACE), and at a community redevelopment organization called the Dudley Street Neighborhood Initiative (DSNI). In Barcelona, I spent time at a local community center and urban garden, and in Havana I volunteered at the Casa del Nino y de la Nina, a youth community center and recreational area.

Finally, I collected data from a variety of secondary sources in each city. The goal was to identify official, media, and other publicly influential perspectives for each neighborhood and to understand their ideal visions for the city and the neighborhood. I also looked for city maps, urban plans, local government legislation, and newspaper editorials and feature articles on the redevelopment of each neighborhood and on local community struggles. Last, I collected publications, reports, and records that were produced by organizations that were working on environmental and health issues in each neighborhood. Through them, I hoped to understand the origins of their work, its evolution, and the strategies and visions that they developed over time and presented to the city.

The questions, research design, and data collection in the study reflect a scholarly concern for highlighting the active residents, workers, and leaders of traditionally distressed urban neighborhoods as well as their supporters—individuals and groups that are not traditionally the focus of planning practice and scholarship. This book is thus based on the categories of people that I decided to study. It examines their experiences in local environmental revitalization and their interpretations of those experiences—without negating the other interests and forms of engagement that exist in these neighborhoods. This research reflects the world that these residents and their supporters occupy, their ways of making sense of their experiences in the city, their thoughts on the decay and historic exclusion, and their strategies for confronting such processes through environmental revitalization endeavors.

The leaders, workers, activists, and supporters that I focus on in this study are not necessarily the majority of residents in each neighborhood or in the city, but they often are the most active (although unheard about in planning and scholarly studies). Because this research is qualitative, interviewees are not meant to be part of a representative sample of a neighborhood. Even so, they are representative of the active participants and leaders of neighborhood struggles and action. By highlighting their battles and their interpretations of their work, I have been able to acknowledge voices that usually are absent from planning practice and general environmental policy and planning studies. I also wanted to avoid reproducing the unevenly balanced power dynamic with which marginalized neighborhoods and cities work. Emphasizing their interpretations and experiences of their neighborhood also explains their long-term commitment to improvement.

My interest in understanding how active residents, leaders, and community workers transformed their neighborhoods required me to account for the types of outside or inside forces, actors, and threats that activists work with, negotiate with, and push back against. I also was committed to understanding the strategies and tactics that activists developed over time and the involvement of their internal and external supporters. This study thus is anchored in the neighborhood residents, leaders, and workers themselves and also accounts for the nonprofit organizations, foundations, and municipal agencies that affected and had a stake in their work.

Book Structure

In the next chapter, I review in greater depth the role that environmental justice studies have played in unraveling the environmental inequalities faced by low-income populations and communities of color (including contamination, resource extraction, and waste transfer) as well as the tensions and conflicts that have arisen from them. I then examine the multifaceted roots of environmental injustices. This book is framed within the political economy of development in the city so that I can explore the broader processes and actors that played central roles in the long-term environmental decay of marginalized neighborhoods and that residents and their allies resisted. Research in sociospatial segregation, inequalities, urban growth, right to the city, and spatial justice is particularly relevant for these issues. In the second part of chapter 2, the traditional understanding of environmental justice is linked with scholarship on place and place attachment and its analysis of the role that place and community play in historically distressed neighborhoods. Here, I critique existing studies of community organization for their lack of attention to place within the context of long-term environmental revitalization and justice in urban distressed neighborhoods, and highlight the need to strengthen the nexus of environmental justice and place remaking in cities.

Chapter 3 presents the baseline conditions of historical marginalization and environmental degradation that activists in Cayo Hueso (Havana), Casc Antic (Barcelona), and Dudley (Boston) faced. It outlines the history of each neighborhood and the forces that contributed to the degradation of the neighborhood and its socioenvironmental and health conditions.

Unlike most environmental justice studies, I examine long-term degradation and changes in each neighborhood and plant the seed for analyzing the relationship between neighborhood decline and local activism and place remaking. Traditional environmental justice studies do not focus on these urban processes and therefore fail to deliver a comprehensive and in-depth view of environmental degradation and reconstruction in the city. Chapter 3 thus lays the groundwork for understanding the connections, relationships, and memories that residents and their supporters have built with their place.

Local activists refer to multiple events, actors, and forces that affected the stability and well-being of the neighborhood and triggered their reactions and mobilization. These include private and public disinvestment, changes in the structure of the local economy and industry, lack of municipal attention to degradation, laissez-faire environmental and land-use policies, redlining and inequitable lending practices, racism, and discrimination against residents. They are also inscribed in a broader context of municipal neglect toward historical and inner-city neighborhoods and of urban renewal and building demolition policies. The multiple causes of degradation and abandonment are structural, collective, and individual. Long-term decay stems from decisions that are made by municipal authorities and urban planners, national programs, neighborhood groups, and individual residents. In Dudley, Cayo Antic, and Cayo Hueso, however, residents who faced neighborhood degradation and long-term marginalization did not remain inactive but instead transformed their neighborhoods. Chapter 3 presents in greater detail the initiatives that residents and their allies organized to transform their neighborhoods. Among others, these include farms, gardens, and schoolyards in Boston, large green public spaces in Barcelona, and green and art projects in Havana.

As has been shown in recent decades, community-based revitalization of distressed and marginalized neighborhoods requires a commitment to the spaces where low-income and minority residents live, learn, work, and play together. Because of a history of community dismantlement and abandonment, activists have faced complex obstacles to transforming their neighborhoods. In chapter 4, I show how activists have been involved in complementary domains that naturally strengthen each other and help achieve community reconstruction.

Indeed, environmental justice requires a consideration of land use, economic development, and environmental protection. Marginalized communities that fight for environmental justice in cities have to go beyond struggles against clearly identifiable "brown" contamination sources. Environmental initiatives are more holistic than they have been traditionally presented. Activists do not compartmentalize their work into discrete categories, such as open space, parks, housing, jobs, and food. They might work on environmental clean-up and safe farming, move to green spaces and physical activity, and then work on learning and education in newly rebuilt spaces. They are interested in both outdoor and indoor habitats. These are the tangible dimensions of place-based urban environmental justice. Urban environmental justice is part of a broader puzzle that involves equitable and sustainable community development and community rebuilding projects. These might be multipurpose community centers (particularly important when outdoor spaces are unused or unsafe), healthy and affordable housing, welcoming venues for healthy food and community activities, and economic opportunities and security through these projects. Community development thus also becomes a tool for advancing environmental justice.

Community efforts to revitalize neighborhoods have been successful thanks to the broad, diverse, and often unexpected coalitions of residents, community organizations, architects, artists, funders, political leaders, and environmental NGOs that have worked within the existing political context. Coalitions have offered residents access to material resources, technical expertise, and financial support. In Dudley, Casc Antic, and Cayo Hueso, this bottom-to-bottom networking was seen in three complementary levels of activism—street activism, technical activism, and funder activism. Coalition members often shared similar interests in neighborhood reconstruction and community power and common values of solidarity, altruism, sharing, and defense of traditionally forgotten residents. Coalitions were diverse because they needed to work on many interrelated issues to achieve long-term environmental justice. At times, loose partnerships grew tighter at crucial moments of neighborhood organization in which common interests in specific projects took precedence over differences in organizing traditions or long-term agendas. For instance, groups with a long-term affordable housing agenda agreed to support the informal self-construction of new parks and playgrounds to make the neighborhood more welcoming and strengthen community ties.

In return, socioenvironmental endeavors and the narratives that activists built around their projects heal grief, decrease loss and violence, create safe havens and refuges, celebrate the neighborhood, and ultimately remake a place and home for residents. This analysis is the focus of chapter 5. Space is a constitutive element of collective action. It is not simply in the background. At the scale of cities, differences in levels of urbanization or political contexts do not substantially affect the experiences and visions of activists for their neighborhood. The stories of Dudley, Cayo Hueso, and Casc Antic activists expose similar experiences of marginalization, abandonment, and grief, together with comparable visions for place remaking. Place plays a dual role as a motivator for action and a goal to achieve. It also has a dual meaning related to both grief and hope.

People feel strongly attached to the places where they live and work, to their tangible assets, and to the relationships that they have built within those places—whether they have been long-term residents or community workers or whether they have recently arrived. These strong attachments have motivated them to engage in environmental revitalization initiatives. Even so, residents in Dudley, Cayo Hueso, and Casc Antic have suffered from environmental trauma and loss, so when activists repair community spaces, build new parks and playgrounds, and develop urban farms and community gardens, they do so to address residents' grief and fear of erasure from a neighborhood that has been seen as a devastated war zone. These activists feel a strong sense of responsibility toward the neighborhood and grow personally as they participate in neighborhood revitalization. Over time, environmental projects help to heal the community, achieve environmental recovery, and create a sense of rootedness and home. They create safe havens for individuals and families, offer a soothing refuge away from the pressures of the city relations, and bolster residents' ability to deal with negative processes and dynamics, especially racist and discriminatory discourses. Activists' visions for place remaking also encompass aspects of safety that go beyond individual protection against physical, social, and financial harm to include soothing, nurturing, and resilience.

As they remake a place for residents, local activists work to create a self-sustained urban village, celebrate the community, rebuild a local collective identity, enhance ties between residents, and encourage them to continue to participate in the reconstruction and protection of their community. Place remaking is thus a dynamic and dialectic activity. Urban sustainability is present in environmental justice activism, but it is enriched here with social dimensions that are not limited to poverty alleviation, local wealth generation, and job creation. Social aspects of urban sustainability include focusing on place remaking, addressing trauma and fear of erasure, and rebuilding a stronger local identity.

In chapter 6, I analyze the broader political goals that activists and their supporters advance when they organize for environmental revitalization.

There is much to learn about how historically excluded groups frame a larger political vision as they understand, resist, and challenge their marginality, especially when they do so as part of long-term environmental revitalization. They challenge public officials and planners who prioritize developments in the neighborhood and decide the neighborhood's significance in Barcelona, Havana, and Boston. They also fight existing racist and classist stigmas and stereotypes about low-income and minority residents—especially that they live in worthless neighborhoods and do not care about the long-term well-being and environmental quality of their place. Most activists Cayo Hueso, Casc Antic, and Dudley note that they could have left and moved to different communities but decided to stay.

Much of the broader political work of these activists has focused on giving residents a greater sense of dignity, addressing vulnerable individual and family situations in their neighborhood, and resisting broader processes of encroachment, excessive tourism development, and environmental gentrification. Beyond claiming a right to the city, they claim a right to their neighborhood. They emphasize that unless they acknowledge development, growth, and gentrification, fighting for community reconstruction and place remaking will be fruitless, and environmental justice achievements will be jeopardized. In some cases, enhancing environmental conditions can have negative consequences on community stability that are potentially greater than locally unwanted land uses (such as toxic sites). When newly rebuilt neighborhoods become desirable to new social and ethnic groups, existing residents may find themselves being pushed away.

Consequently, activists in Dudley, Casc Antic, and Cayo Hueso have aimed at controlling the projects and activities that take place within the neighborhood territory, gaining secure tenure over the land, and building residents' stewardship of it. They have set up and maintained physical and symbolic borders with outsiders whom they see as threats to the stability and cohesion of their neighborhood. New environmental spaces (such as playgrounds, urban farms, and sports centers) are sometimes borders themselves, creating a clear separation between us and them and discouraging visitors from entering the neighborhood or appropriating community gardens. They suggest that those projects belong to residents—and not to outsiders.

As activists fight against outside threats and influences, they create self-managed spaces and new models for democratic planning and participation in the city. This involves transgressing existing norms, asking for greater autonomous and spontaneous management of the neighborhood (especially of public spaces) and of the environmental initiatives they fostered, which they consider to be learning spaces and commons to protect. In Barcelona and Boston, residents and their supporters have questioned traditional democratic institutions and processes, and in Havana, they called for less bureaucratic decision making and planning. In other words, community struggles over environmental and health conditions in impoverished neighborhoods have constituted the basis for questioning political and institutional power systems in the city, recapturing the roots of democracy (Purcell 2008), and allowing communities of color and low-income neighborhood to have a greater say in processes that affect their neighborhood, even if tensions and disagreements remain within the neighborhood.

Finally, chapter 7, the conclusion, proposes a new framework for understanding environmental justice action in cities. The right to a healthy environment requires certain physical and psychological dimensions of community rebuilding and place remaking. For residents, this right cannot be separated from working on the land, border control, community engagement, and a renewed form of local planning and democratic practice. In some ways, this framework is not far from radical utopian urban planning procedures. All of its elements need to be in place to build just and revitalized cities. The planning and policy initiatives for sustained urban revitalization and environmental justice must avoid perpetrating new processes of mental and physical displacement for vulnerable urban residents.

2

Environmental Justice, Urban Development, and Place Identity

Environmental justice scholarship and studies on place and place attachment in historically marginalized neighborhoods are vast and well developed, and I will not review them in depth here. In this chapter, I connect existing knowledge on environmental inequalities to the structural mechanisms that produce them and to a political economy of environmental discrimination and unequal urban development. Environmental inequalities are situated within broader processes of urban growth and often originate in policy and planning decisions that negatively affect residents and the stability and quality of their neighborhoods. I present North American, international, and transnational perspectives on environmental discrimination and inequities, urban development, and community struggles. I also look at how studies on environmental justice and movements enhance an understanding of neighborhoods as places toward which historically marginalized groups develop strong attachments, build durable ties, and assign specific meanings.

Traditional Perspectives on Environmental Inequalities

Unequal Exposure, Burden, and Access
In the early 1980s in the United States, poor citizens and communities of color were harmed by neighborhood dumping and contamination and outraged by their lack of voice in environmental policy implementation. They initiated street protests in the south. Since then, an extensive literature in sociology, environmental policy, and environmental health has examined inequities in health and exposure to contaminants. Hundreds of reports and articles have shown that minorities and low-income populations have suffered greater environmental harm than white and wealthier communities from waste sites, incinerators, refineries, transportation, and specific area sources (Bryant and Mohai 1992; Bullard 1990; Pellow

2000; Varga, Kiss, and Ember 2002; Downey and Hawkins 2008; Lerner 2005; Sze 2007; Mitchell and Dorling 2003). Exposure to harm and risk is not limited to the place where they live. Poor working conditions also often require farm and factory workers to be in close contact with pesticides and hazardous waste (Harrison 2011; Pellow 2002; Pellow and Park 2002; Smith, Sonnenfeld, and Pellow 2006). In the Silicon Valley south of San Francisco, California, for instance, Asian and Latino female workers have had to manipulate highly toxic substances like xylene and glycol ethers in the manufacturing of microelectronics (Pellow and Park, 2002).

Such inequities are exacerbated by the fact that communities of color and low-income neighborhoods have traditionally received less environmental protection than more privileged areas. Governmental policy making, regulating capacity, and oversight of contaminating industries have traditionally been weak (Pellow 2001; Checker 2008; Morello-Frosch 2002). Laissez-faire policies have accompanied lenient corporate practices in the areas of safety, toxic release, and waste management. Thirty-five years after toxic waste was discovered to have contaminated the Love Canal neighborhood of Niagara Falls, New York (Blum 2008), schools and houses still exist close to a gas refinery in neighborhoods inhabited by poor and immigrant residents. Children breathe toxic fumes as they play soccer a few hundred meters away from the plant. In other areas, residential houses are situated next to lots where asbestos-covered construction materials have been piled.

The term *environmental racism* refers to race-based decisions to site hazardous facilities or to remediate environmental hazards in the United States. As a result, populations of color suffer from intentional or unintentional harm when projects and policies negatively affect their health and environment (Bryant and Mohai 1992). In the United States, for instance, locally unwanted land uses (LULUs) have traditionally been sited in poor black and Latino neighborhoods rather than in affluent suburbs (Bullard 1990; Pellow 2000; Schlosberg 2007; Corburn 2005; Blodgett 2006). For example, in Los Angeles in the early 1980s, more than 60 percent of Hispanics and only 35 percent of whites in Los Angeles lived in waste-site areas (Bullard 1990). In Warren County, North Carolina, in 1982, a local landfill of toxic polychlorinated biphenyls (PCBs) leached into the drinking water supply of the town's black residents. Thirty years later, statistics are still disheartening. A recent study of the Bronx reveals that people who live next to noxious land uses are 30 percent more likely to be poor and 13 percent more likely to be members of a minority

(Maantay 2007). Even high-income minorities are directly or indirectly the target of contaminating facilities.

In the global north, environmental inequalities among poor and minority populations are not unique to the United States. Central and eastern Europe's environmentally hazardous activities (such as the illegal disassembly of car batteries from dumps and gas stations) are disproportionally located in areas with a high concentration of ethnic and national minorities, such as Romany (gypsy) populations (Varga, Kiss, and Ember 2002). Indigenous and minority populations in the republics of the former Soviet Union are also affected by offshore oil drilling, natural resource extraction in land reserves, limited access to rural livelihoods, and instability of the legal environment (Agyeman and Ogneva-Himmelberger 2009). Likewise, in the autonomous community of Catalonia in Spain, contaminating industrial facilities (such as metal smelting factories, chemical industries, and waste-management facilities) tend to overburden low-income neighborhoods outside cities such as Barcelona and Tarragona (Ortega Cerdà and Calaf Forn 2010). Studies of outdoor air quality in northern Europe demonstrate a similar relationship among socioeconomic status, air quality, and combined effects on respiratory health (Mitchell and Dorling 2003; Wheeler and Ben-Shlomo 2005). In the United Kingdom, for instance, households from lower socioeconomic classes are more likely than those from higher-income groups to live in areas of poor air quality.

Inequalities exist in the distribution of environmental harms and also in the territorial allocation of environmental goods and services. Low-income urban neighborhoods tend to receive poor environmental services (such as street cleaning and waste collection), and wealthy communities enjoy environmental privileges (parks, coasts, open space, and sports centers)—often in a racially exclusive way (Pellow 2009; Landry and Chakraborty 2009; Heynen, Perkins, and Roy 2006; Hastings 2007; Carruthers 2008; Estabrooks, Lee, and Gyurcsik 2003; Lovasi, Hutson, Guerra, and Neckerman 2009; Park and Pellow 2011). In many cases, wealthier and white groups enjoy parks and forests thanks to the work done by immigrant and poor workers, whom they accuse of damaging natural resources and attempt to exclude from spaces of environmental privilege (Park and Pellow 2011). An environmental justice concern is not merely the number of parks and green spaces and access to them but their safety. In distressed communities, the few parks are often unsafe or used for drug dealing and consuming. They are not always well maintained, and parents prefer to keep their children away from them.

In many cities, deprived neighborhoods are often food deserts or fast-food jungles with few supermarkets and produce markets and many fast-food restaurants and convenience stores. Food deserts are prevalent in the United States and Canada, but they also are found in European cities, especially in the United Kingdom (Wrigley, Warm, Margetts, and Lowe 2004; Smoyer-Tomic, Spence, and Amrhein 2006; Beaulac, Kristjansson, and Cummins 2009; Guy, Graham, and Heather 2004). The term *food desert* reportedly originated in Scotland in the early 1990s to illustrate the difficulties that poor urban residents encountered when they tried to access a diet that was affordable and healthy (Cummins and Macintyre 2002). Since the early 2000s, several studies have provided statistical evidence on food inequities in cities. Today, white neighborhoods in U.S. cities have on average four times more supermarkets than black neighborhoods have (Morland, Wing, Roux, and Poole 2002). In states such as Maryland, New York, and North Carolina, wealthy neighborhoods have twice as many supermarkets as low-income neighborhoods have (Moore and Diez Roux 2006). Lower-income families have been shown to travel on average between 1 and 1.6 miles to do their food shopping (Hillier, Cannuscio, Karpyn, McLaughlin, Chilton, and Glanz 2011), which requires them to own a car if public transit is poor (which is often the case). Local scarcity of healthy food is thus exacerbated by low numbers of car ownership and a lack of access to efficient public transportation that would allow residents to shop in more distant grocery stores and markets. Lower-income residents also tend to work two or three jobs and have little time to shop for food in distant locations.

In a similar way, in the global south, environmental issues (most often in the form of hazards, contamination, and intensive resource extraction) tend to disproportionally affect poor and minority populations. Over the past decades, mercury spills from gold mines, oil and timber extraction, deforestation and erosion from widespread farming, and giant hydroelectric dams have devastated millions of hectares in Latin America, Asia, and Africa (Brysk 2000; Carruthers 2008; Evans, Goodman, and Lansbury 2002; Hilson 2002; Martínez Alier 2002; Ahmad 1999; Carmin and Agyeman 2011). In the Amazon region of South America and in Ogoniland in Nigeria, the activities of oil and mining companies have contaminated indigenous and farming communities, leading to the disruption of nutrition patterns and to increased incidences of cancer (Bastos et al. 2006; Wheatley and Wheatley 2000; San Sebastián, Armstrong, Córdoba, and Stephens 2001). Researchers and activists have started talking about hydric or water justice to highlight inequities in access to and use of water resources (Boelens, Cremers, and Zwarteveen 2011). For instance, urban residents in Cochabamba, Bolivia, have lost water access when water delivery and sanitation became privatized, and the Embera Katío indigenous people and fishermen in Colombia have seen their sources of livelihoods become endangered by the construction of the Urrá dam and paramilitary control of their land.

These patterns are intensified by governments that generally are less able and willing to regulate transnational industries and by weak or poorly monitored corporate social responsibility business schemes (Blowfield and Frynas 2005; Frynas 2005; Vogel 2006; Carmin and Agyeman 2011). Community-engagement initiatives of corporations often fail because the affected people are not truly involved, questions of legitimacy and representation are not addressed, and development programs are missing (Blowfield and Frynas 2005; Frynas 2005; Newell 2005). For instance, businesses that attempt to improve their practices often do not identify local populations that are affected by investment practices as legitimate stakeholders who have the right to articulate concerns (Newell 2005).

In addition to material extraction or contamination of land and water resources, millions of tons of toxic waste from industry, agriculture, consumers, public institutions, and computer and electronic products are exported every year to poor countries (Martínez Alier 2002; Pellow 2007). This waste contributes to elevated rates of human mortality and ecosystem damage (Pellow 2007). Global environmental change involves complex and multiscale environmental inequalities. Climate changes exacerbate inequalities between the global north and south and reinforce existing injustices within countries. Many of the countries that are facing rising oceans, increased droughts, and extreme disasters are the least responsible for climatic changes and have the fewest resources for coping with resultant challenges (Parks and Roberts 2006; Satterthwaite, Huq, Reid, Pelling, and Romero-Lankao 2007).

Within countries, poor residents (who tend to live in areas that are exposed to unstable climate conditions, such as floodplains and coastlines) are often the most vulnerable to extreme weather and landslides, floods, and infrastructure damage (Anguelovski and Roberts 2011; Parks and Roberts 2006; Adger 2006). Mega cities such as Mumbai or Jakarta present drastic inequalities—between lush secluded communities that benefit from environmental amenities and unauthorized slums that are not connected to city services such as waste collection or water provision and whose often migrated to these areas to escape war or poverty in rural regions.

The experiences of historically distressed communities indicate a clear and pervasive relationship between environmental inequalities and health (Corburn 2005; Shruder-Frechette 2007). Air and water contamination is directly related to public health problems such as respiratory diseases, infectious diseases, and cancers (Brulle and Pellow 2006). The scarcity of parks, playgrounds, fitness clubs, community centers, and other physical activity facilities is linked to subsequent disparities in health-related behaviors and obesity in individuals (Estabrooks, Lee, and Gyurcsik 2003; Lovasi et al. 2009). Similar relationships exist between health problems and an inequitable distribution of grocery stores and fresh food options. In food deserts, poor and minority communities have fewer healthy food choices than nonminority and richer neighborhoods do and are closely associated with higher rates of obesity, cardiovascular disease, and atherosclerosis (Morland, Wing, and Roux 2002; Dunn 2010). More often than not, multiple health problems exist concurrently.

Community Voices against Environmental Inequalities

Residents of distressed communities have not remained silent when faced with environmental inequalities. Initially, environmental justice coalitions in the United States organized to protest against contaminating facilities and highway construction in residential areas (Chambers 2007; Pellow and Brulle 2005; McGurty 2007). Since the 1980s, there have been hundreds of environmental justice protests, and many have resulted in victories for distressed communities. In 2001, residents of Anniston, Alabama, won a $42.8 million settlement against Monsanto for polluting their city with PCBs over several decades, and they also obtained the relocation of affected residents. Even when U.S. courts remain silent, some communities have sought other ways to obtain justice. In 2005, African American residents in Mossville, Louisiana, objected to the industrial pollution produced by fourteen chemical plants and filed a complaint against the plants before the Inter-American Commission of Human Rights. Over the past three decades, companies have had to improve their emissions standards, pay financial compensations to affected families, and comply with U.S. Environmental Protection Agency (EPA) requirements to control noxious activities. Although in 2005 the U.S. General Accounting Office released a report showing that the EPA devoted little attention to environmental justice issues between 2000 and 2004 while drafting three significant clean-air rules on gasoline, diesel, and ozone, several landmark victories have sent a signal to contaminating industries that they cannot continue doing business as usual.

In the United States, many early environmental justice activists were active in the 1960s civil rights movement and later expanded their approach to reflect a broader human-rights perspective (Bullard 2005; Pellow and Brulle 2005; McGurty 2007) and a gender perspective (Stein 2004; Pellow and Park 2002; Kurtz 2007). Labor activists work to ensure that businesses provide safer conditions for workers who manipulate contaminated materials and waste (Harrison 2011; Pellow 2002; Pellow and Park 2002; Smith, Sonnenfeld, and Pellow 2006).Today the concept of environment is defined more broadly than it was in earlier years. Environment is considered to be the place where people live, work, learn, and play together. For those in the environmental justice movement, every person, regardless of race, income, culture, and gender, has the right to a decent quality of life (Gauna 2008). This new definition of environment as a safe place to live, work, learn, and play came out of the 1991 First National People of Color Environmental Leadership Summit, and it has become widely used by public health and planning experts when they examine the relationship between health and place.[1]

Early environmental justice activists positioned themselves outside the conventional environmental movement. This was particularly true in the United States, where they accused traditional environmental NGOs of reifying pristine and wild ecosystems (Bullard 1990; Gauna 2008; Schlosberg 2007; Shutkin 2000; Dobson 1998; Pulido 1996; Sandler and Pezzullo 2007). Conservation organizations such as the Sierra Club and the World Wildlife Fund (WWF) were often viewed as prioritizing the protection of nature and parks above people's health, jobs, well-being, and identity. They were accused of being complicit in the conservation policies that government agencies imposed to control the land and its people and modernize the country in many places around the world (Robbins 2011). For instance, traditional environmentalists might push to create protected forest reserves while minimizing the reliance of indigenous communities on the forest for food and claiming that parks must be guarded from human presence (Lowrey 2008). In many cases, local stakeholders who have a lay knowledge of sustainable natural management can be excluded from the protected areas. In other words, activists in the environmental justice movement have denounced programs that leave vulnerable communities without access to traditional resources, jobs, and livelihoods.

By including resources, jobs and socioeconomic well-being in its goals, the environmental justice movement links environmentalism to a broader quest for social and economic justice.[2] For environmental justice activists, because justice and sustainability cannot be separated from each other, greening and environmental projects must be envisioned with social equity and wealth creation in mind (Dobson 1998; Pulido 1996; Agyeman, Bullard, and Evans 2003). Over the years, the environmental justice agenda has expanded to include the right to well-connected, affordable, and clean transit systems in cities (Agyeman and Evans 2003; Loh and Eng 2010; Loh and Sugerman-Brozan 2002; Lucas 2004). For instance, in the Los Angeles, California, area, the Transit Riders for Public Transportation (TRPT) campaign has contested structural inequities in transportation funding and investment in the area, which prioritizes suburban expansion over the preservation of the existing system, and has asked that a Title 6 provision be established to prohibit racial discrimination in federally funded transit projects.

In addition, the right to healthy, fresh, local, and affordable food for community food security is at the center of much community advocacy (Gottlieb 2005, 2009; Alkon and Agyeman 2011; Gottlieb and Joshi 2010; Hess 2009). Urban farms are sprouting to address food deserts and make culturally valued food available to communities of color. On the ground, many projects address not only the need for food security but also food sovereignty—that is, the control over the means and processes of food production. Through farming, residents have access to new healthy food sources and build closer social ties with each other. Last, in the housing domain, community organizations advocate for recycling and playground spaces within public housing, and for green, affordable, and healthy housing (i.e., Loh and Eng 2010).

Environmental justice groups also have adjusted to the economic downturn and reorganized to seize opportunities arising from new policies. In the past few years, the green economy has become a priority (Fitzgerald 2010), and interest has grown in job training for energy-efficiency projects and funding from utility companies for weatherizing houses (Fitzgerald 2010). Residents and community organizations associate initiatives for enhancing housing stock quality and increasing revenues for low-income and minority communities with initiatives for reducing carbon dioxide emissions and addressing climate change. In Massachusetts, the Green Justice Coalition's energy-efficiency campaign helps obtain small subsidies that allow low- and moderate-income residents to weatherize their houses. Environmentalism here connects social equity and new forms of wealth creation to sustainability (Agyeman and Evans 2003) and to climate mitigation. In shrinking and declining cities that have been affected by the economic crisis and accelerated deindustrialization, residents are coming together to merge empty lots and transform them into parks, playgrounds, and farms (Dewar and Linn, forthcoming).

In the global south, activists since the early 1990s have demanded greater local participation in matters of social justice, indigenous peoples' rights, access to land, labor rights, wealth redistribution, and opportunities for engaged participation in land-use decisions (i.e., Carruthers 2008; Martínez Alier 2001, 2002; Evans et al. 2002; Newell 2005; Doyle and Risely 2008). Environmental abuses have triggered numerous violent and nonviolent conflicts in which residents organize against the private appropriation of communal livelihoods and the structures of economic and political power (Martínez Alier 2002; Pellow 2007; Shiva and Bedi 2002). In India, poor communities have used nonviolent protests to resist the privatization of natural resources such as water and land and the violation of human rights, including those of women (Shiva 2002; Shiva and Bandyopadhyay 1991). Latin America also has a long history of environmental justice struggles, mostly with mining and oil extraction in indigenous communities and more recently with trade-related land struggles (Carruthers 2008).

In the south, the "environmentalism of the poor" is embodied in environmental-distribution conflicts (Martínez Alier 2002). This form of environmentalism in mainly interested "not in a sacred reverence for Nature, but a material interest in the environment as a source and a requirement for livelihood; not so much a concern with the rights of other species and of future generations of humans as a concern for today's poor humans" (Martínez Alier 2002, 11). According to the "environmentalism of the poor," many poor people generally do not feel, think, and behave as environmentalists do and may mistreat the environment. Nevertheless, early studies of conflicts in India and Latin America revealed that that because the poor rely directly on the land and its natural resources and services, they often are strongly motivated to be careful managers of the environment (Guha 1989, 2000; Martínez Alier 1991; Davey 2009). Today many local communities organize to protest against the behavior of private corporations because they have a vital interest in the natural resources around them and because they assign specific cultural and social values to those resources. Their valuation is different from the valuation of private corporate actors (Martinez-Alier 2009).

In urban cities in the south, there has been only limited scholarly attention paid to community organization for greater environmental quality and access to environmental benefits in vulnerable neighborhoods. A greater emphasis is put on rural areas or on issues of waste and waste manipulation in cities (Agyeman et al. 2003; Carruthers 2008; Pellow 2007). Some scholars of urban livability have emphasized the importance of decent livelihoods for ordinary urban residents in developing cities. For these scholars, livable cities are places that offer affordable housing, decent infrastructure, and healthy habitats and that maintain the ecology of the space (Evans 2002). However, such studies leave little room for an analysis of agency among distressed communities and of neighborhood-based mobilization for enhancing local environmental conditions. The notion of justice and rights is also absent. Yet, in cities such as Bangalore or Mexico, many residents mount resistance to protest airports and highways or gated communities because they affect their territory. Others, such as the Alliance of Indian Wastepickers (AIW), organize to secure a living collecting, sorting, recycling, and selling materials that individuals and industries have discarded, and they protest modern incinerators that would take away their main source of income.

Political Economy of Environmental Injustices in the City and Urban Conflicts

The Multifaceted Roots of Environmental Injustices
The causes of environmental injustices are complex and interlocking and include a lack of recognition of identity and difference between groups and individuals, a lack of attention to the social context in which unjust distribution takes place, and an unequal access to decision-making processes (Schlosberg 2007). People are not always able to persuade decision makers to consider their individual needs, priorities, and capacities. Inequalities and injustices also stem from stakeholders (such as the state, community development organizations, and private firms) with often contradictory and shifting interests and allegiances who struggle for access to scarce resources, including clean and safe living, recreational and work spaces (Pellow 2000). In other words, multiple structures of domination create and reproduce environmental injustices and discriminatory practices (Pellow and Brulle 2005; Young 1990; Pellow 2000).

Within urban distressed neighborhoods, because racial segregation in housing has established the conditions for neighborhood decay, devaluation, disinvestment, and environmental burden, socioeconomic factors and the contextual and policy dimensions of health inequities need to be examined (Frieden 2010; McClintock 2011). For instance, communities of color are often disproportionately exposed to environmental hazards because of social processes such as racial discrimination in housing rentals and sales, zoning laws, redlining by insurance companies and lending institutions, and neighborhood covenants which prevent them to move easily through the urban space (Maantay 2002; Self 2003; Sugrue 2005). In deindustrializing cities like Detroit, the Home Owners' Loan Corporation and the Federal Home Loan Bank used to give white neighborhoods higher marks for loan agreements if they had racial covenants protecting residents from the "infiltration of inharmonious racial or nationality groups," as recommended by the Federal Housing Administration back in 1938 (Sugrue 2005; Hillier 2003). Geographers also point to the power of real estate developers to invest in, disinvest from, and reshape neighborhoods (Harvey 1989). Individual and structural racism has thus contributed to the creation of ghettos and spatial segregation in cities.

Throughout the 1970s and 1980s, higher-income residents fled the center of many American cities for the suburbs, lured by new and affordable homeownership possibilities and new jobs and industries. The growth of suburban sprawl was paralleled by job loss, neighborhood decline, and housing decay and demolition for poorer and minority residents of the inner city (Massey and Denton 1993). A city like Oakland, California, exemplifies this process. Between the 1960s and the 1970s, 130 factories closed in Oakland, many new industries emerged in the suburbs, property owners stopped maintaining housing structures due to redlining and a poor climate for investment, and the construction of new highways (such as the Cypress Freeway) razed homes and displaced thousands of low-income Oaklanders (McClintock 2011). This combination of industrial redevelopment, economic restructuring, insufficient public resources for inner-city areas, and urban renewal projects throughout the United States left many neighborhoods half abandoned and dilapidated and residents with few economic opportunities, thereby exacerbating existing racial tensions (Massey and Denton 1993). In a few years, neighborhoods turned from stable and welcoming to grim and undesirable, opening the door for an unequal siting of environmental harms and an unequal application of environmental laws.

Patterns of racial segregation produced by planning decisions and zoning regulations are less common in Europe (Agyeman 2002). However, as global and entrepreneurial cities such as London and Barcelona compete to position themselves favorably to attract private resources, local governments are less willing or able to regulate capital investment, and they exclude some neighborhoods and social groups from the benefits of new development (Brenner 2009). Traditionally, the image of the growth machine has been used to describe the behavior of elites, rentiers, and their economic and political coalitions as the motor of unregulated growth and private capital accumulation (Logan and Molotch 1987; Squires 1991; Harvey 2003). Because investments move from place to place in cycles of growth, devaluation, destruction, reinvestment, and mobilization, development ends up being spatially uneven throughout the city (Harvey 1981; Smith 1982). Some areas benefit from a flow of public and private resources while others are scarred by abandonment and flight.

Today, as large cities become integrated within global flows, the image of nodes often is used to describe a broad system of complex socioeconomic forces. Local actors are caught within a network of financial services and corporate headquarters that organize the internationalization of production, finance, and information (Friedmann 1986; Sassen 1999; Abrahamson 2004). This transformation of cities has been paralleled by growing socioeconomic inequality between workers who rely on manual skills and workers with higher education and technological skills (O'Connor 2001). The power of poor workers has been declining, as capitalist accumulation geographically disperses investments away from politicized center locations to minimize labor costs (Harvey 1996; Merrifield and Swyngedouw 1997; Soja 2000). As a result, social redistribution, community cohesion, and a strong local government that protects the rights of vulnerable groups have become jeopardized. Inner cities become attractive to high-income professionals who can afford new condominiums, expensive shops and restaurants, and high-priced cultural events, while low-income and migrant residents either move away or are confined to substandard and cramped housing in degraded districts of Paris, Barcelona, and London (Borja, Castella, Belil, and Benner et al. 1997; Borja, Muxí, and Cenicacelaya 2004).

Part of this process reflects the production side of gentrification, through which private developers, investors, and individuals from privileged backgrounds buy the property of less well-off families in poor and working-class inner-city areas (Anderson 1990; Smith 1986). As Smith argues, this forms the basis of what he calls the "rent gap"—the difference between "the actual capitalized ground rent (land value) of a plot of land given its present use and the potential ground rent that might be gleaned under a 'higher and better' use" (Smith 1987, 462). Through the rent gaps, potential profits can be made by reinvesting in degraded and abandoned properties that can be inhabited by new residents who have the money to purchase or rent them.

In large urban areas, some correlation seems to exist between urban land clean-up and the processes of gentrification. As a formerly depressed and marginalized area benefits from new investment from public and private resources, neighborhood indicators start to change. Land and real estate economists have shown that the clean-up of Superfund sites has been related to a significant appreciation in housing values—up to 18.5 percent for blocks within 1 kilometer of the site (Gamper-Rabindran, Mastromonaco, and Timmins 2011). So-called deprived neighborhoods also can experience residential changes through greening and sustainability projects. This process is not neutral and displaces residents who often fought for the improvements. It has been called *ecological gentrification*—that is "the implementation of an environmental planning agenda related to public green spaces that leads to the displacement or exclusion of the most economically vulnerable human population while espousing an environmental ethic" (Dooling 2009, 630). New or restored environmental goods tend to be accompanied by rising property values, which in turn attracts wealthier groups and creates greater gaps with poorer neighborhoods (Gould and Lewis 2012). The city of Portland, Oregon, has promoted neighborhood livability initiatives while preserving industrial details. As the project creates narratives of restored ecologies and an optimistic future, it also sells the new revitalized urban landscapes to wealthier residents who can bring new capital and consumption into the local economy (Hagerman 2007).

In the global south, inequalities in toxic exposure and resource extraction require a political economy lens that accounts for different scales and stakeholders. Several scholars have used a transnational life-cycle approach to consumption, production, and hazards, examining the extraction of materials and the exportation of environmental bads across spaces and stages. In the environmental justice literature, this process is known as the *treadmill of production* (Pellow 2000; Schnaiberg, Pellow, and Weinberg 2002; Schnaiberg and Gould 1994; Schnaiberg 1980). As progress in technology drives the expansion of production and consumption in a synergetic way and as states, investors, and workers depend on economic growth to achieve job creation and revenues, various cycles (of unstopped production, extraction of material and natural resources, and waste accumulation) reproduce over time. Because countries in the south do not have the ability or will to impose their own rules over transnational industries (Newell 2001; Vogel 2006), poorer and more excluded communities suffer disproportionately. This is true in both cities and rural areas.

As in cities in the north, transnational flows of investment, information, and production are transforming urban spaces in rapidly urbanizing southern metropolises. Developers and real estate actors apply pressure on urban lands and territories. In Mumbai, "entrepreneurs of space" replaced the slums of Dharavi with multistory condominiums (Roy 2011), and development mafias often partnered with real estate capital, state actors, and the police (Weinstein 2008). In Shanghai, millions of urban residents have been forced off their land and removed from their networks to allow for the construction of buildings and infrastructure with cutting-edge designs (Wu 2000). The familiar processes of urban restructuring also are found in postsocialist cities and growing cities in the global south, usually with the complicity of national and local governments. Cities' expansions toward the peripheries have often translated into taking over the land of urban and peri-urban farmers (Goldman and Longhofer 2009). Some megaprojects such as airports or financial centers have led to mass inequality and displacement (Golderman and Longhofer 2009; Wu 2004). Cities have grown in size, become more dense, urbanized, and less homogenous. New forms of urban poverty have appeared out of the economic restructuring and reduction of the size of the labor force in state enterprises.

As a result, cities like Lima, Sao Paolo, and Johannesburg have great disparities between rich, gated, and well-maintained neighborhoods and informal settlements where rural migrants and low-income families live without regular potable water or electricity. The distance between the rich and poor and between whites and people of color is nonnegotiable. Even in gray spaces, the wealthy own expensive real estate, and the poor are seen as homeless and landless invaders who have to live in substandard housing. Here, the state often formalizes and criminalizes different forms of spatial occupations and configurations, transforming gray spaces into black spaces (Yiftachel 2009).

Yet residents do not remain inactive when they are displaced from their land so that their neighborhoods can be transformed into world cities and globalized locales searching for new frontiers of urban development. Many protest new airports, highways, gated communities, and medical tourism facilities as in Bangalore and Mexico City (Goldman and Longhofer 2009; Davis and Rosan 2004). In Bangalore in 2008, for example, hundreds of residents protested the felling of 90,000 trees for the construction of a new elevated Metro system (Goldman and Longhofer 2009).

A Broader Interpretation of Urban Struggles and Movements

The causes of environmental injustices are deeply rooted and multifaceted. Because of the complexity of the processes that create inequalities between groups and neighborhoods, broad frameworks are used to analyze and interpret urban environmental conflicts. These struggles are not just against the outcomes of unfair and racist decisions and practices or the unequal applications of environmental protection laws.

Some theorists say that conflicts reflect the underlying tensions between place entrepreneurs and labor. In the past, social movement scholars felt that urban movements expressed structural contradictions in the city and that they potentially could bring about radical changes in political power (Castells 1977). As workers resisted the degradation and disruption of their neighborhoods, they began to protest capitalist accumulations on their space (Harvey 1981; Smith 1982; Mollenkopf 1981). Their struggle combined socioeconomic demands with collective interests (especially collective consumption) in a struggle for community culture and political self-determination (Castells 1983). However, the geographical dispersal of investments away from politicized center locations has weakened workers' place and power (Harvey 1996; Merrifield and Swyngedouw 1997; Soja 2000; Castells 1983), and urban social movements became either coopted by private interests or repressed by the state (Castells 1983). Macroforces transformed the power of urban social movements from proactive to reactive.

More recently, theorists of urban movements have developed broader and more inclusive frameworks that acknowledge that activism has moved from factories to neighborhood streets and public spaces. New social movements view cities and local spaces as objects of protest and push for social changes in culture, identity, and urban development. Under slogans such as "Another New York Is Possible," demonstrators contest growth politics, corporate urban development and entrepreneurialism in the city, commercialization of the public space, and neglect of neighborhoods. Scholars of social theory and critical geography have advanced three frameworks for describing these movements—the just city, spatial justice, and the right to the city. Other frameworks, such as the revanchist city and the creative city, have not received the same level of attention.

Susan Fainstein (1999, 2006, 2011) has been the most vocal defender of the just-city framework. For her, achieving a just city requires creating different values of democracy, equity, growth, diversity, and sustainability; implementing processes of participation, contestation, and democratic planning; and pursing concrete outcomes. In some ways, residents' claims resonate with the utopian idea of the good city—a city with a self-organizing and active civil society that resists within democratic institutions and that establishes the minimal political, economic, social, and ecological conditions that are necessary for communities to thrive (Friedmann 2000). A good city honors the rights of every resident to flourish.

Spatial justice is another framework that can provide an overarching explanation and rallying point for social and environmental struggles. Some view urban claims as spatial-justice demands (Soja 2009). Political theorists and urban geographers define *spatial justice* as the equitable allocation of socially valued resources in space (including jobs, political power, social status, income, social services, and environmental goods) and equitable opportunities to make use of these resources (Marcuse 2009a, 2009b, 2009c; Soja 2009). They consider spatial justice as the broader dimension from which other demands for equity, including environmental justice, can and should be derived. For Marcuse and Soja, attention must be paid to the spatial dimensions of urban inequalities and to the active forces that create social inequalities in a space if one wants to comprehend demands for environmental justice. Combining space and justice opens up new opportunities for the analysis of urban movements.

A dialectical dynamic is at the center of spatial injustices: social and human processes shape spatial patterns, as much as spatial patterns shape social processes (Marcuse 2009a, 2009b, 2009c; Soja 2009). Examples include the redlining of urban investments, residential segregation, and open-space planning (Soja 2009; Fainstein 2011). Spatial injustices are viewed as being constructed in a space that systematically creates oppression, and inequities are seen as the consequence of spatial-domination practices (Merrifield and Swyngedouw 1997). As Dikeç argues (2001, 179), "spatial dynamics do not only aggravate but actually produce injustice through the stabilization of social inequalities and problems, becoming a major reproducer of them." Achieving spatial justice today requires progressive and participatory forms of democratic politics through grassroot social activism (Soja 2010) and a rebuilding of the urban common by recapturing the land and its assets and creating new forms of governance (Chatterton 2010).

Finally, the concept of a right to the city provides a helpful lens through which to analyze activists' struggles for environmental quality and livability. Theory building around the right to the city is not new. Several decades ago, the French geographer Henri Lefebvre argued that this right is earned by taking part in the daily making of the urban fabric by living in the city and by meeting particular responsibilities that entitle people to participate in decision making (Lefebvre 1968, 1972; Lefebvre, Kofman, and Lebas 1996; Mitchell 2003). People control spaces of production and also use and shape the city. Cities are not meant to be for profit and developers but for the people themselves (Mayer 2009, 2012, 367): "The right to the city is less a juridical right, but rather an oppositional demand, which challenges the claims of the rich and powerful. It is a right to redistribution, as Peter Marcuse once called it, not for all humans, but for those deprived of it and in need of it. And it is a right that exists only as people appropriate it (and the city). It is this revolutionary form of appropriation, which Lefebvre meant to discover in 1968 Paris." From a procedural standpoint, the right to the city is about democratizing planning and political practices in cities and affecting decision-making processes.

Today, urban activists use the idea of the right to the city, and coalitions such as the Right to the City Alliance[3] have emerged in the United States. In addition to demands for economic and environmental justice, coalition members ask for greater democracy and the end of real estate speculation, community space privatization, and gentrification (Steil and Connolly 2009; Marcuse 2009a). Through their local engagement, activists attempt to implement new complex and radical sociospatial initiatives that question the commodification of the urban space and put in place self-management and self-organization projects (de Souza 2006; Montagna 2006). In East Harlem, New York, recent fights for decent housing have been transformed into a struggle for self-determination and against neoliberalism, hypercommodification of the urban space, gentrification, and displacement (Maeckelbergh 2012).

The movement for a right to the city has traveled beyond the United States and into eastern and western Europe, including Hamburg, Berlin, and Zagreb, where activists worked against the gentrification of Flower Square and its surrounding area (Čaldarović and Šarinić 2009). In the Occupy movement in the United States and the Movimiento 15-M in Spain, people mobilized around public squares, against foreclosed housing, for a different form of democracy, and against the domination of developers and bankers in urban economies. In the global south, transnational urban movements have emerged. To this date, however, scholars have paid little attention to urban movements (such as those that were featured at the 2009 World Social Forum) or to public debates on the right to the city in developing cities. For instance, in 2005, the United Nations Human Settlements Program (UN-HABITAT) and the United Nations Educatioinal, Scientific, and Cultural Organization (UNESCO) initiated a public debate through the Urban Policies and the Right to the City group to identify best practices and find synergies that result in equal access to the benefits of the city for all urban dwellers, while promoting sustainable urbanization.

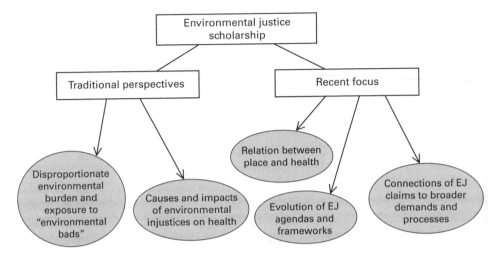

Figure 2.1

An overview of environmental justice scholarship

In sum, the mobilization of marginalized residents occurs in spaces that have experienced numerous transformations, disruptions, and fragmentations. Over time, such processes have produced environmental inequalities and given rise to numerous local conflicts that have been studied and interpreted through a variety of lenses. Figure 2.1 summarizes the extent of environmental justice scholarship today.

Urban neighborhoods are not neutral and empty recipients of the results of residents' struggles against environmental inequalities. Cities are important convergence and rally points for marginalized residents and their supporters as they fight for long-term environmental quality. Residents develop connections to their neighborhood, have built numerous ties and relationships within it, and display strong feelings and emotions when they relate their neighborhood experiences. Their struggles are grounded in a space and a place that in turn affect their engagement and participation, as I examine in greater depth below in the last section of this chapter.

Place and Community Engagement in Urban Distressed Neighborhoods

The Meaning of Place and Attachment for Marginalized Residents

Researchers in sociology, planning, geography, and environmental psychology have examined and theorized about place in cities. In this book,

I offer a glimpse into this vast and multidisciplinary scholarship and concentrate on how previous studies help us understand the relationships among place attachment, sense of community, and participation in urban neighborhoods, especially historically disenfranchised neighborhoods.

A *place* is a meaningful site that combines location (the geographical situation of a place), locale (its physical and material characteristics), and sense of place (the feelings and emotions that are induced by a place, especially through representation) (Cresswell 2009). Local social relations and their links to a broader system produce a place over time (Massey 1994). Places are imbued with layers of sedimented meaning that are derived from tradition, identification, and sentiment (Corcoran 2002; Hayden 1995) and from the experiences and social connections that people have built there (Cresswell 2009). The relationship between people and places is reciprocal (Cummins, Curtis, Diez-Roux, and Macintyre 2007). People shape places through their interactions with each other and with their locale, and places shape residents through the political and physical processes that intervene in their lives. On the one hand, adolescents playing soccer tournaments on empty lots in La Mina, a Barcelona neighborhood on the outskirts of the city, create a livelier and warmer feel to a grim neighborhood and inspire families to use public places as their own. On the other hand, past experiences of muggings on the streets and of fear of police interventions discourage some families from allowing their children to play outside. They prefer to keep them indoors.

Place attachment is an essential characteristic of people's feeling toward a place, and it is defined as the affective bond between people and places (Low and Altman 1992). Ordinary people extract from spaces a clearly identified, bounded, and meaningful place (De Certeau 1984) with which they build a strong connection. Place attachment provides a sense of security and well-being, defines boundaries between groups, and anchors memories, especially against the passage of time (Gieryn 2000; Logan and Molotch 1987). It can rest on physical features and settings (such as the urban built environment) as well as social dimensions (Scannell and Gifford 2010). The ties that exist among residents are based on connections between the people who live there and also on the setting, building layout, and characteristics of public or semipublic spaces. Narrow streets and alleys allow people to develop social and cultural outdoor activities so that residents recreate a piece of village life and build connections to each other (Gans 1962; Small 2004). Such a duality is illustrated in scholarly debates between environmental determinism views and social compositionalist views on community and place.

The attachment that people feel toward a place is often connected to their sense of community, including the feelings that they have about belonging to a group (Doolittle and MacDonald 1978). It includes a connection to a shared history and concerns (Perkins and Long 2002) and is shaped and strengthened through relationships and experiences. For instance, long-term residents in a place that is undergoing transformation sometimes feel emotionally rooted to their place and nostalgic about memories that include children's games played on the street, informal conversations at bakeries, and solidarity among neighbors during hardships (Corcoran 2002).

Essential to the understanding of place attachment is its connection to questions of identity. The relationships of people to a place and the feelings that they develop toward it contribute to the formation and protection of their identity (Altman, Low, and Maretzki 1992; Low and Lawrence-Zúñiga 2003). Place identity is shaped through interactions that create values and beliefs (Fredrickson and Anderson 1999; Bondi 1993). Because of the importance of experience and time, place identity is intimately connected to memory (Hayden 1995; Bondi 1993). If people lose their memories of a place, they lose their identity. Residents build their own memories of a place during war, oppression, and conflicts, and these memories can be very different from representations of the same place by public officials (Tills 2005). In Argentina, for instance, the Madres de la Plaza de Mayo spent their lives looking for traces of their kidnapped children while the military dictatorship denied that they did the kidnappings and did everything possible to hide any trace or evidence of those kidnappings. As Leonie Sandercok (1998, 207) writes, "memory locates us, as part of a family history, as part of a tribe or community, as part of city-building and nation-making." In other words, the identity of a place is often the object of controversies, conflicts, and traumas.

Historically marginalized neighborhoods tend to be perceived by both outsiders and insiders as poor urban ghettos that are scarred by violence, dilapidation, and poverty. Others consider them to be isolated and bounded neighborhoods—badlands or social purgatories where only urban outcasts dwell. In France, they are the *banlieues*; in Lima, the *pueblos jóvenes*; in Boston, Mattapan; in northern Paris, the Seine-Saint-Denis department (department 93, le Quatre-vingt treize); in Rio, Mandela. Prejudicial beliefs are attached to these areas, which in turn set up further socially negative effects for their residents (Wacquant 2007; Garbin and Millington 2012). The labels and the public policies that are applied in these areas often reinforce their marginality and limit the creation of new

In the United States, the stigmatization of neighborhoods been used to justify criticism against public housing, ıewal projects, and disperse poor residents, especially go and New York, without considering the intimate rks that residents have built there (Crump 2003). This l stigmatization eventually dissolves the value of the ıd discourages them from seeing and using the neighborhood as a resource (Wacquant 2007).

As much o urban sociology literature argues, the neighborhood ; a critically imp nt place for marginalized groups. Concepts such as place attachment sense of community help us understand how impoverished and min ity residents feel about their place, what uses they have developed for it, and what meanings they assign to them (Clark 1989; Manzo 2003; McAuley 1998; Pattillo 2007; Falk 2004). In many cases, working-class and minority neighborhoods are much more than what the media describe as urban ghettos scarred by violence and poverty. The close-knit families who live there value community life, social ties, and their roots in the neighborhood. Residents come to rely on each other and build bonds of mutual support within, between, and across neighborhood spaces. The experiences of residents in HOPE VI public housing revitalization projects in the United States reveal the presence of well-functioning networks and numerous self-help practices (Manzo, Kleit, and Couch 2008).

In schools, churches, barbershops, and cafés, residents create casual and impromptu connections and find comfort and social reinforcement for their beliefs (Gregory 1998; May 2001; Anderson 2003). In the United States, black churches and taverns are essential resources and places for people to find consistent support, learn to deal with outside relationships, and understand the roles played by outside influences in their lives (May 2001). Residents that spend ample time together display resourcefulness in moments of hardship and community-building (Small 2004; Falk 2004). Two dimensions of attachment are striking at the neighborhood scale—bondedness (a feeling of being part of a neighborhood) and rootedness to a community (Riger and Lavrakas 1981).

In the environmental arena, neighborhood green space plays multiple roles for low-income populations and residents of color. Residents use public spaces with plants and trees to develop social contacts with each other and feel less vulnerable in their neighborhood (Kuo, Sullivan, Coley, and Brunson 1998). Vegetation supports the use of common space and encourages neighbors to build social contacts with each other.

Urban community gardens have been shown to play multiple roles for low-income populations and residents of color. Gardens provide people with educational activities, cultivate a self-help ethos, promote gardeners' pride, and create strong bonds with the space (Lawson 2005; Schmelz-kopf 2002). Gardening has important social and mental benefits because it allows people to relax, teach others, learn, and connect with others. Regular contact with nature can help prevent and treat mental health problems (Maller, Townsend, Pryor, Brown, and St. Leger 2006; Marcus and Barnes 1999). Gardens help people release stress, escape from conflict, restore hope, and improve their sense of well-being (Francis and Hester 1999; Marcus and Barnes 1999). In other words, they heal and are restorative (Gerlach-Spriggs, Kaufman, and Warner 2004).

Most neighborhoods go through phases of evolution and change, but the stability of urban distressed neighborhoods is particularly threatened by disruption and fragmentation, negative labels, urban-renewal projects, and gentrification (Small 2004; Gans 1962; Pattillo 2007; Smith 1982; Fullilove 2004; Otero 2010). Developers, investors, and individuals from privileged backgrounds have been shown to buy property in poor and working-class inner-cities areas, and those residents are pushed away directly or indirectly (Anderson 1990; Smith 1986). In the United States, for thirty years, urban-renewal policies have promoted redevelopment projects that have triggered feelings of amputation as new private housing replaces public housing and as overlapping networks built in corner stores, gardens, and street activities are destroyed (Fullilove 2004; Otero 2010).

The loss of place has devastating consequences for protecting individual and collective memory and identity as well as for restoring mental wellness (Fullilove 1996). When modernist planners displace people and demolish houses, they leave residents homeless physically and mentally as their memories are destroyed (Sandercock 2003). They put red dots on the neighborhood and implement policies that transform its urban structure and social fabric. As projects develop, newcomers settle in the neighborhood, replace former residents, erase their common practices, and change the existing identity of the place.

Contesting and Organizing in the Neighborhood

When vulnerable urban residents face the disruption and fragmentation of their place, they often do not remain passive or silent. In cities, daily forms of subtle contestation disavow and reconstruct place identity. First, residents often contest the stigmas that are associated with degradation and neglect and instead try to create meaningful and autonomous images of place and community (Gotham and Brumley 2002; Falk 2004). They might claim that their place has more liveliness, diversity, and cultural life than the rest of the city in which they live. In some instances, such as the Cité des 4000 in La Courneuve next to Paris, young people participate in art projects that foster creativity and revise the negative stigmas associated with the neighborhood (Garbin and Millington 2012). As fragile neighborhoods undergo gentrification, immigrant residents use the oral and written discourse in theater plays to create social and moral boundaries for newcomers (Molan 2007). Their artistic practices delineate borders and group membership and notify newcomers that their presence is not desired and that they are occupying a space that is not theirs.

Second, residents of impoverished neighborhoods often transgress and resist what policy makers, planners, and media sources consider to be the normal use of a place. They confront critiques about so-called undesirable behaviors, such as hanging out outside, dancing on the streets and sidewalks, talking loudly, drawing graffiti, and playing music from balconies. By engaging in activities that create and sustain a new and positive personal identity that is tied to a place (Gotham 2003), they contest traditional social constructions of their place and construct autonomous behavior. The activities in which they engage help them create and sustain an identity that is connected to place (Gotham 2003). As they hang out at street corners or play dominos on benches, they demonstrate their belonging to their neighborhood and their natural use of the public space.

Third, beyond the creation of autonomous images and experiences of places, residents take actions to protect the neighborhood. When people are attached to their place, they tend to invest time in it, monitor local development, participate in community planning, and unify others around them (Chavis and Wandersman 1990; Cohrun 1994). Such behavior illustrates the idea of Gemeinschaft (Tönnies 1957): individuals' collective sentiments and holistic social relationships drive them to be more oriented toward the large collective association rather than their own self. They seek not their private benefit but rather the improvement and well-being of the larger group to which they belong. Residents with greater emotional connections to their place will be more likely to engage in neighborhood participation, collective action, and eventually political action (Gotham and Brumley 2002; Davis 1991; Tilly 1974; Suttles 1968). They interact more with their neighbors, invest their time, watch over developments in the neighborhoods, and unify others around them (Brown, Perkins, and Brown 2004). They also tend to feel more obliged to participate in the processes that affect their neighborhood (Chavis and Wandersman 1990; Cohrun 1994; Davidson and Cotter 1993).

In urban distressed communities, resident participation is often rooted in domestic property relations. People fight against threats to land, houses, amenities, and safety (Cox 1982; Gotham and Brumley 2002; Davis 1991; Venkatesh 2000); they defend their housing complexes, poverty-reduction policies, and social welfare programs (Fisher 1984; Pattillo 2007; Peterman 2000; Sampson 2004; Von Hoffman 2003); and they promote gardening on their block or in their building (Hou 2010; Lawson 2005). When a drug economy emerges and safety becomes a concern, the neighborhood loses its identity, character, and tradition. It becomes disembedded. As acts of resistance, residents attend public hearings, organize protests and rallies at traditional landmarks within the neighborhood, call for community planning meetings, and lobby their local representative or congressman. In resisting disruption and threats, people develop coping mechanisms and tactics that they can use to oppose oppressive transformations and precariousness and differentiate themselves from the changes.

Residents' actions are counterpoints to claims that capitalism erases the accumulation of shared history, collective memory, and neighborliness of strangers (Sennett 2007). Despite living in cities where capitalist investments and processes outmaneuver urban movements, residents preserve a strong attachment to their place, resist changes, and do not renounce their neighborhoods. Their behavior also undermines claims that civic togetherness (daily community-based engagement in practices such as local organizations and meetings) has disappeared and led to a decline of social capital and community (Putnam 2001). Such claims have called attention to the excessive individualism and separation between people in today's cities but do not hold true everywhere and in all contexts.

Toward New Environmental Justice Studies

Until recently and despite growing research on the environment in the United States (Agyeman et al. 2003; Gottlieb 2005, 2009; Pellow and Brulle 2005; Checker 2011; Gould and Lewis 2012; Diaz 2005), most academic work examining environmental inequalities has focused on environmental threats and hazards (the "brown" EJ) to the health and livelihoods of marginalized communities internationally. Numerous studies exist on the siting of landfills and refineries, on health issues related to toxic waste sites and improper waste management, and on environmental accidents and contamination as part of resource extraction processes (the most notable manifestation of this emphasis is the journal *Environmental Justice* and the numerous studies published on environmental burdens and harmful policies and projects).[4] Furthermore, much of the environmental justice research in cities has examined pollution without giving enough attention to the everyday characteristics of urban landscapes, such as the mundane but chronic manifestations of socioecological injustice in urban spaces (Bickerstaff, Bulkeley, and Painter 2009), and to the broader urban transformations that occur in neighborhoods and cities. We know very little about community organization around initiatives that contribute to holistic and long-term neighborhood transformation. Studies of individual themes (such as transportation justice and food justice) are more readily available. More attention should be given to activism, engagement, and the underlying meanings of local struggles. Especially needed are comparative and transnational perspectives on these issues.

Second, the term *environment* generally refers to the natural features of a place and to its air, water, plants, and soil. Researchers generally predefine the environment of people and places and environmental justice action. In this book, I question established understandings and boundaries of environmental justice. I challenge traditional representations of environment in the lives of low-income and minority residents and in the ways that people define and view their environment. I also examine the complex and interlocking dimensions of health in urban environmental revitalization while keeping in mind the environmental justice movement's definition of *environment* as being a place where people live, work, and play together.

Third and most important, little is known about how place, place attachment, and sense of community shape struggles for long-term environmental quality in historically deprived neighborhoods. Few urban studies have examined the sense of place and identity that connects activists who fight for environmental improvements.[5] Empirical work is needed to investigate how activists experience a neighborhood that has undergone long-term environmental degradation and decay (not necessarily from a toxic waste site or factory) and how activists assign meanings to their environmental revitalization work.

In this book, I integrate existing knowledge about place in cities into the scholarship on environmental justice to examine the role that is played by place attachment and identity in local environmental engagement. I seek to understand the complex underlying agendas of activists who create and advocate for socioenvironmental projects within a broader urban context. What forms of connections do activists develop over time and express toward marginalized neighborhoods with long-term substandard environmental conditions? How are questions of place and people's sense of community expressed in local environmental justice struggles? How do residents' experiences of their neighborhood shape their engagement in environmental-revitalization projects? What motivates residents to take action in their neighborhood and remain committed to it despite its dire baseline conditions and the numerous obstacles to transforming it?

The actions of individuals and groups have a broad effect on their neighborhood. As activists shape their place through their daily behaviors, they indirectly or directly formulate and negotiate other claims. One of my goals is to unravel how community and place are being reshaped through local environmental mobilization. This means examining how neighborhood groups improve environmental conditions and at the same time question, realign, and recreate identities in the process of gathering support. I also assess here whether claims to spatial justice and a right to the city are adequate frameworks for understanding activists' work on the ground. By examining urban case studies that represent a well-established democracy, a young democracy, and an autocratic regime, I compare whether and how differences in political systems and urban development influence activists' claims and community organization, as well as their interpretation of the processes that threaten their neighborhood.

The next chapter introduces the baseline conditions of marginalization and environmental degradation that were faced by residents in Casc Antic (Barcelona), Cayo Hueso (Havana), and Dudley (Boston). The chapter is structured around the history of each neighborhood and serves as a reference point for examining the challenges of each place and for later analyzing the role of memory, identity, abandonment, and loss in community organization around environmental revitalization work. As I developed my field research, I became aware of the importance of the past and of community life in activists' positions and justifications for neighborhood engagement. External forces have resulted in neighborhood landscapes that include patches of empty lots, semiabandoned houses, substandard infrastructure and services, and closed storefronts. People's memories and past experiences are constituted by specific events and threats as well as by long processes of decay, unequal development, and hope of revival.

3

Stories of Neighborhood Abandonment, Degradation, and Transformation

In the Dudley, Cayo Hueso, and Casc Antic neighborhoods, activists struggled within complex historical processes of marginalization as well as racial, social, and spatial inequities. In each of these three neighborhoods, residents and their supporters responded to a past history of uneven development, overall degradation and abandonment, and substandard environmental and health conditions. The Dudley neighborhood in Boston faced decades of disinvestment, white flight, and racial violence. Casc Antic in Barcelona experienced periods of abandonment but more recently unequal social and environmental revitalization as the new democratic municipality of Barcelona tried to redevelop the city after the dictatorship. Finally, Cayo Hueso in Havana has been the site of gradual degradation, first under U.S. occupation and later under the Castro regime. Such neighborhood conditions constituted the baseline from which environmental-improvement initiatives emerged at the end of the 1980s and beginning of the 1990s. Activists refer to processes of abandonment and degradation in their testimonies about community organizing and their engagement in the neighborhood.

In this chapter, I draw on many sources—reports, newspaper articles, planning documents, and interviews with leaders, residents, community workers, nongovernmental organizations, foundations, planners, and decision makers—to describe the past abandonment and marginalization of Dudley, Casc Antic, and Cayo Hueso in Boston, Barcelona, and Havana. I also describe the environmental and health improvements that residents, leaders, and community organizations achieved in their neighborhoods through a variety of projects.

From Prosperity to Decay, Wasteland, and Revival in the Dudley
Neighborhood in Boston

Historical Dynamics of Exclusion and Marginalization

*White Flight, Slow Degradation, and Urban Renewal in the 1950s,
1960s, and 1970s*   In the 1940s, Dudley was a thriving district within
Roxbury, one of Boston's twenty-one official neighborhoods. The streets
around Upham's Corner, which was at the intersection of five streetcar
lines, were flourishing. Two business areas bustled with people who vis-
ited restaurants, shops, and cafés. Every week, Saint Patrick's church at
the corner of Dudley Street and Magazine Street organized social events
that helped newly arrived Roman Catholic Italian and Irish immigrants
integrate themselves into the neighborhood (Warner 1978). Long-time
residents still remember that era with fondness and melancholy. By the
late 1950s, however, white residents started to move away from Boston
and out to the city's prosperous and modern suburbs in search of more
space and better schools, services, and infrastructure (Levine and Harmon
1992). The white population in the Dudley neighborhood dropped from
95 percent in 1950 to 45 percent in 1970. The houses that white middle-
class owners left behind were in substandard condition, as demonstrated
by faulty plumbing, outdated heating systems, and facades with peeling
paint (Rosenthal 1976).

   As Dudley's social characteristics changed, so did economic and business opportunities. In the 1950s, the mechanization of agriculture and
mining extraction in the southern states of the United States left millions
of workers jobless and prompted many families to migrate north. Between
1940 and 1960, the African American population in Boston increased
from 20,000 to 60,000 (Lukas 1985), and by the 1970s, black residents
accounted for 53 percent of Roxbury's population. Furthermore, during
the 1970s immigrants from Puerto Rico, the Dominican Republic, Hon-
duras, Guatemala, and Cuba expanded the Latino population in Dudley,
which reached 28 percent in 1980 (Medoff and Sklar 1994). As newcomers arrived, older residents continued to move out, and rents started to
fall. In three decades, the white exodus to the Boston suburbs emptied
the city of nearly 30 percent of its population, and the commercial sector
around Dudley Street (mostly banks, small businesses, restaurants, and
factories) started to decline, leaving Dudley without its engine of economic development. Manufacturing jobs decreased from 20,000 in 1947 to
4,000 in 1981, the number of businesses on Dudley Street fell from 210 in

1950 to 74 in the 1970s, and the rate of unemployment among Dudley's residents rose to twice the rate of other sections of Boston (Medoff and Sklar 1994). By the end of the 1970s, poverty rates in Roxbury had sky-rocketed, and 32 percent of households (especially black families) were living in poverty (Cotton 1992).

In the 1970s, in addition to disinvestment and abandonment, property neglect started to disfigure Dudley's landscape. Absentee landlords rented their properties to newcomers, and did not maintain or improve the condition of their properties. For property owners, it was often easier to deal with Boston building inspectors and pay fines than with banks to obtain loans for property renovation (Levine and Harmon 1992). As landlords left properties abandoned or semiabandoned, tax revenues fell, and infrastructure maintenance and municipal services suffered cuts (Warner and Durlach 1987). White residents lived in modern suburbs around Boston, but Roxbury's residents of color had inferior living conditions with minimal municipal intervention.

Discriminatory attitudes were particularly evident when black or Latino families sought to rent or buy a house. As in many other communities, banks considered any area with a large number of minority residents to be in decline and therefore not worth investing in (Putnam, Feldstein, and Cohen 2004; Sugrue 2005). In addition to this redlining, arson was so frequent in Dudley that children awoke every night to the sounds of firefighters' sirens and the screams of people escaping flames, but the police did not investigate arson properly. It also allowed dealers to sell drugs on the streets in front of residents' homes without any fear of arrest. By the 1980s, various socioeconomic, political, and individual decisions had transformed the Dudley neighborhood from a quiet place for working-class families to a segregated ghetto where people were afraid to talk to each other and felt unsafe waiting at bus stops.

At that time, the city of Boston took advantage of the passage of federal urban-renewal legislation to clear degraded areas through the city. First, in the 1950s, the municipality allowed the destruction of many blocks in the West End neighborhood between the North End and Beacon Hill, which displaced 2,600 families, and replaced low-rise working-class apartment buildings with expensive high-rise apartment towers (Green 1986). This destruction reflected the vision of a new Boston that was put forth by the Boston Redevelopment Authority (BRA): it envisioned a revitalized downtown that was close to the financial district and that included profitable, tax-yielding, centrally located real estate (Levine and Harmon 1992). Other similar controversial projects did not achieve the desired objectives of planners and developers. When the city of Boston and the BRA planned to tear down a block of low-income housing in the South End and to replace it with parking spaces for nearby office workers in 1968, black and white residents formed a coalition to protest the city-sponsored plan (King 1981). The official strategy was controversial because the BRA had no plan to build low-income housing in the South End despite the increasing need for healthy and quality homes for poor residents (Warner and Durlach 1987). After days of protests, sit-ins, nights spent in tents, and police arrests, residents won the battle, and the BRA developed plans for housing construction.

During that period, protesters against urban renewal often collaborated with activists working in other movements, such as the antihighway movement. In Boston, the antihighway movement emerged at the end of the 1960s when new plans for highway construction were disclosed. Designs included the construction of an elevated highway on top of cleared land in Roxbury and Jamaica Plain. To oppose this highway project, a wide coalition of academics, local activists, and politicians in Roxbury and Jamaica Plain led protests against the highway and in 1970 persuaded Massachusetts Governor Francis Sargent to halt construction. Forty years later, new construction projects broke ground in 2012 as part of the Jackson Square revitalization, but much of the land cleared for the new highway in Roxbury and Jamaica Plain remains vacant. The highway was not built, and the battles over its construction continue to have a lasting effect on the local landscape and on the residents. Yet those early victories energized local movements in Boston, and many activists moved on to address violence and develop community gardening in Boston's marginalized neighborhoods, including Dudley (Warner and Durlach 1987).

Violence, Desegregation, and Racism in the 1970s and 1980s   In the early 1970s, in the aftermath of the civil rights struggles in the U.S. South, protests erupted in Boston against discriminatory practices, especially in schools. In Roxbury, the conditions in black neighborhood schools were substandard: classrooms had broken windows and furniture, students read from old battered textbooks, and teachers sometimes used corporal punishment on students (Lukas 1985). When a U.S. District Court judge ordered desegregation measures to be imposed on Boston public schools in 1974, the school system started using buses to transport white and black students back and forth from their home neighborhoods to schools in other neighborhoods to create mixed-race classrooms (Warner and Durlach 1987). When these desegregation measures were implemented, white residents in the Irish American neighborhood of South Boston protested the busing of black students from Roxbury and Dorchester into South Boston and the removal of their students to other neighborhoods. When in 1974, South Boston residents threw stones at the buses transporting black students, nine black students were injured as white families shouted racial slurs at them and tried to reclaim the sovereignty of their neighborhood (MacDonald 1999). There were house arsons and murders. Concerns became more acute when the white political leaders in the city of Boston failed to address racial violence.

At the time of desegregation and school busing in Boston, a new coalition of political activists emerged. Mel King's 1979 mayoral campaign brought together people from different Boston neighborhoods who tackled violence, access to employment, wealth distribution, taxes, and representation. Interest groups such as the Alliance for Rent Control helped to develop position papers reflecting proposed solutions. Eventually, the campaign led to the creation of a Boston-wide, neighborhood-based, multiple-issue organization, the Boston People's Organization, which sought to gain control of neighborhoods, jobs, and governments and to empower oppressed residents (King 1981). Its work marked a new stage in the development of a progressive political force in the city. As Mel King pushed people to take responsibility for their own community and eliminate social negatives, discrimination, and racism in neighborhoods of color, he also sought to decentralize problem solving and tackling environmental issues. His political program focused on housing code enforcement (for unsafe and unsanitary housing), garbage on street and alleys, air pollution, open-space development, recreation, alternative energy sources, and gardening (King 1981).

This environmental emphasis reflected the further decline of the Dorchester and Roxbury neighborhoods in the 1970s and early 1980s. In Dudley, the arrival of working-class residents of color and their demand for housing was coupled with new waves of arson—by developers who were eager to push out low-income populations, gut rehab buildings, and sell them and by white property owners who were eager to collect insurance money and buy new property in the suburbs (Layzer 2006). Between 1947 and 1976, half of Dudley's housing stock (648 buildings) was destroyed (Shutkin 2000). Between 1970 and 1980, white flight to the suburbs left behind interracial violence, abandoned neighborhoods, and poverty (Warner and Durlach 1987). By 1990, one out of two Dudley children was living below poverty level and was malnourished. The neighborhood also had a disproportionate level of homicides (Medoff and Sklar 1994). Over the years, marginalization, poverty, stereotypes, and prejudice had deeply transformed Dudley—for the worst. Environmental health conditions were particularly alarming.

Environmental Neglect and Health Emergencies

Waste and Dumping   By the early 1980s, after years of abandonment, decay, and arson, Dudley was a dumping ground. A third of the neighborhood's land was vacant—approximately 1,300 lots (or 21 percent of all parcels) in a radius of 1.5 square miles. For years, contractors from outside the city, residents from other neighborhoods, and Boston builders had deposited truckloads of old appliances, rotten meat, automobile bodies and parts, unused construction materials, and trash onto empty lots near residents' homes in Dudley (Medoff and Sklar 1994). This dumping was accompanied by a growth in other illicit activities, such as drug dealing and illegal toxic waste discarding, with offenders taking advantage of poor lighting and lack of police attention around the vacant lots (Settles 1994). Many schoolyards were also dilapidated and underused, reflecting inadequate funding for public schools in low-income neighborhoods and the use of these spaces for criminal activity. Schoolyards were considered wasted urban land rather than physically active environments that had the potential to improve learning and offer greater recreational opportunities for children (Lopez, Campbell, and Jennings 2008).

At that time, changes in local business development also affected Dudley's physical conditions. Auto-repair shops, waste-handling companies, scrap-metal dealers, and truck storage facilities filled the neighborhood and contaminated the land (figure 3.1). Auto-body workshops, gas stations, and storage tanks spread oil-based contaminants underground as by-products of their activity (Shutkin 2000). By the mid-1990s, the Massachusetts Department of Environmental Protection identified fifty-four hazardous waste sites in Dudley, including many contaminated with lead, chromium, mercury, asbestos, and petroleum constituents. Further environmental concerns arose because the sites of industrial businesses were in dense housing developments, private houses, and schools and because their activities were not monitored or regulated.

Additional environmental and health problems arose in the 1980s from the increased presence of trash-transfer stations, many of them illegal and uncontrolled. Their negative effects on the neighborhood often stemmed from the odors, noise, and noxious air emissions they released. In addition to these sources of contamination, residents suffered from

(a)   (b)

Figure 3.1

Dudley: (a) Trash dumping and (b) empty lot contaminated with asbestos

lead poisoning through their contact with contaminated soil and water (Settles 1994). By the end of the 1980s, almost every street in the Dudley area had a child who had been diagnosed with lead poisoning. Ground-water was also contaminated with metals from foundries that had leached into the soil, volatile organic compounds (VOCs) from underground storage tanks, and other toxic substances. With this high prevalence of environmental pollutants, asthma hospitalization rates in Roxbury were 5.5 times the Massachusetts average (Massachusetts Division of Health Care Finance and Policy 1997).

Environmental and health problems in Dudley were not addressed by existing environmental laws and regulations. The Massachusetts Hazardous Waste Management Law (Massachusetts General Laws, chapter 21E) was modeled after the federal Superfund Act, and it made (and still makes) it difficult for affected parties to hold a private company or individual liable. Even though chapter 21E requires the parties responsible for land and groundwater contamination to remediate the affected land and pay in full for clean-up, parties generally have litigated the issue of liability to delay clean-up or simply negate responsibility. Even after a clean-up is completed, companies can still be held responsible to third parties, providing no end point to liability (Shutkin 2000), so in the 1980s and early 1990s, polluters often abandoned sites and moved to cheaper land outside the city. Furthermore, trash-transfer stations and other unwanted land uses fell between the cracks of most environmental laws. Their status as small businesses exempted them from the requirement to provide comprehensive environmental review under the Massachusetts Environmental Policy Act (MEPA). They were considered ordinary land-use decisions by Boston's zoning board and were decided on a case-by-case basis. The board lacked formal environmental review authority. Moreover, until the late 1980s, Boston officials ignored Dudley residents' pleas to address routine dumping and health hazards.

A Food and Recreational Desert   In addition to environmental and health hazards, the Dudley neighborhood lacked major facilities that were common in other areas of Boston, such as supermarkets, pharmacies, community centers, parks, and recreational facilities (Settles 1994) (figure 3.2). Physical inactivity in Roxbury reached 72 percent versus 63 percent in Boston as a whole and 38 percent in high-income neighborhoods due, in great part, to the lack of parks and playgrounds. In the 1980s, most grocery stores closed down and were replaced by fast-food restaurants and convenience stores that offered prepackaged and fried foods. Residents had little time or money to access distant and often expensive healthy food options. They were part of the 2.3 million Americans who lived more than a mile from a supermarket and did not have access to a car (Ver Ploeg, Breneman, Farrigan, Hamrick, Hopkins, Kaufman, Lin, et al. 2009). In 2000, 71 percent of blacks and 69.4 percent of Latinos in Boston still had inadequate vegetable and fruit consumption (Boston Public Health Commission 2004). In 2001, heart disease was the second leading cause of death for Roxbury residents at 213 deaths per 100,000, and obesity was the first key health issue in the community (Boston Public Health Commission 2004). In 2001, 61.8 percent of all Roxbury's black residents and 62.8 percent of Latino residents were considered to be obese. Dudley was a wasteland in multiple ways: it was a food desert, a desert in recreational facilities and park, and a contaminated desert.

Enhanced Environmental Quality in Dudley
On a beautiful summer day in 2013 in the Boston neighborhood of Roxbury, it is hard for visitors to imagine the succession of historical periods that have made it what it is today. Roxbury was once a small colonial farming town, a Boston middle-class suburb, an industrial site patched with immigrant housing, a blighted and arson-ridden postindustrial neighborhood, and now a rising neighborhood.

Contaminated and devastated lots were at the forefront of early restoration efforts in the Dudley section of Roxbury.[1] In 1985, the Dudley Street Neighborhood Initiative (DSNI) was born with a mission of

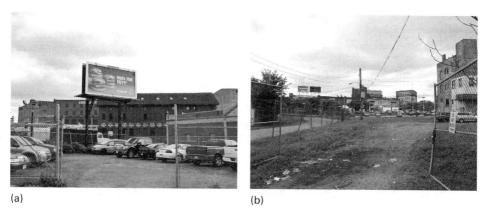

(a)   (b)

Figure 3.2

Dudley: (a) A car and bus depot with a fast-food advertising sign and (b) a desolated area

"empowering Dudley residents to organize, plan for, and create and control a vibrant, high quality, and diverse neighborhood in collaboration with community partners" (Shutkin 2000). DSNI began its mobilization work in 1986 with a comprehensive cleanup campaign called Don't Dump on Us, which created the basis for sustained resident participation in environmental-improvement efforts (Layzer 2006). The campaign centered on cleaning up more than 1,300 lots (fifty of which had been diagnosed as hazardous waste sites). In 1987, DSNI targeted two illegal trash-transfer sites in Dudley and convinced the Boston Public Health Commission to shut them down. After its initial successes, DSNI involved citizens in community planning to achieve broad-based revitalization and community control over the land. At that time, 181 parcels of empty land in Dudley were owned by absentee companies, individuals, or the city of Boston. In 1988, DSNI successfully lobbied the city of Boston to gain eminent domain over a triangle of sixty-four acres of abandoned land (later known as the Dudley Triangle) within its borders (Medoff and Sklar 1994).

In the early 1990s, local nonprofit organizations targeted health hazards such as lead poisoning by combining prevention with clean-up work. In 1994, DSNI and ACE (which provided legal assistance and capacity building to groups that were working on illegal dumping and pollution) became founding members of the Massachusetts Environmental Justice Network and won three legal cases against dirty businesses. ACE also led a comprehensive effort to clean up empty lots. In 1995, it joined DSNI, the Bowdoin Street Health Center, the Environment Defense Fund, and the Massachusetts Campaign to Clean Up Hazardous Waste to create the Neighborhoods against Urban Pollution (NAUP). NAUP was a resident-led organization that identified, mapped, prioritized, and cleaned up environmental hazards, including sites in Dudley. In 1998, The Food Project, an environmental nongovernmental organization that was created in 1991, began working in 160 backyard gardens to educate gardeners about lead in urban soils and safe gardening practices.

When environmental threats became controlled, residents and organizations turned to open-space and park rehabilitation and development. In 1996, DSNI created the Dudley Town Common, which was made up of passive open spaces that residents helped to design. Among other parks, the Dennis Street Park was built in 2009 through two grants from the state Urban Self-Help funds and the City of Boston Capital Improvement Program. The budget for the park was $550,000, and it included a children's water play feature and a play area with equipment (Boston Parks and Recreation Department 2007). In Dudley, the Youth Environmental Network, local professionals and organizations that trained youth residents to help with park maintenance and stewardship, helped coordinate additional park revitalization and clean-up efforts.

Many open-space revitalization efforts in Dudley were associated with environmental education programs and projects that took place in neighborhood parks and schoolyards (figure 3.3). The Food Project partnered with the Shirley-Eustis House Association and built an Urban Learning Farm at the Shirley-Eustis House—the former mansion of colonial Governors Shirley and Eustis and today a national historic landmark. In addition, the Boston Schoolyard Initiative (BSI) has reconstructed abandoned schoolyards with play equipment, games, maps, gardens or natural areas, and an amphitheater or a stage for teachers to conduct classes and children to play spontaneously (Boston Schoolyard Initiative 2009). In Dudley, BSI orchestrated interventions at the Dearborn Middle School, Emerson Elementary School, and Mason Elementary School. In parallel, informal neighborhood associations such as the Upham's Corner Westside Neighborhood Association formed around open-space improvements, ecological restoration, raised beds, rain barrels, and urban sustainability.

In regards to sports and physical activity, residents and community organizations have worked on renovating and creating new sports grounds and community centers with gyms and sports equipment. In its early years, DSNI reclaimed the Mary Hannon Park, which had been taken over by

(a)   (b)

Figure 3.3

Dudley: (a) A new children's tot lot and (b) a recently created schoolyard at the Mason School

drug users. DSNI also coordinated the creation of two new community centers—Orchard Gardens and Vine Street. The community organization Project Right also enhanced physical activity opportunities while helping neighborhood stabilization and economic development (Project Right 2008). For instance, in the Green Space Improvements Program, Project Right coordinated the partial or full renovation of tot lots and playgrounds, such as the Holborn Street tot lot for $375,000 (Project Right 2008). On a smaller scale, the Body by Brandy Fitness Studio founded in 1996 by Brandy Cruthird is one of the first and only gyms in the United States that is owned by black women. Another of her successes is the Body by Brandy 4 Kidz nonprofit program. In 2011, the Kroc Foundation completed the Kroc Center, a $115.5 million dollar aquatic center and fitness facility proposed by community residents who met in meetings and workshops that were convened and led by DSNI.

Early in the revitalization of the Dudley neighborhood, community leaders, organizations, and environmental NGOs concentrated on developing urban farming. Urban agriculture and community gardens were developed after World War II in the Fenway area of Boston as part of the Frederick Law Olmsted's Emerald Necklace park system and later through the work of organizations such as Boston Urban Gardeners (BUG) and the Boston Natural Areas Network (BNAN).[2] In Dudley, local gardeners took the lead in the early 1990s in cleaning empty lots, dividing them into plots for residents, and organizing farming and social activities inside the gardens. Over the years, gardeners have received regular technical guidance from BNAN, which today supports thirteen community gardens in Dudley. Within DSNI, Greg Watson and Trish Settles helped launch the Urban Agriculture Strategy, which cleaned up brownfields and vacant parcels to redevelop them for food production. Beginning in 1997, with support from the U.S. Environmental Protection Agency, the Massachusetts Highway Department, a $172,000 grant from the Ford Foundation, and $60,000 from the Noyes Foundation, several brownfield sites received environmental remediation to become community gardens. DSNI also built a 10,000 square foot greenhouse.

Today, Dudley is home to three urban farms run by The Food Project (figure 3.4). The organization started its lots in 1991 and received technical and legal support from ACE on land-transfer issues. Every year, The Food Project grows over 250,000 pounds of chemical-pesticide-free food for charitable donation and youth-driven food enterprises. The Food Project is best known for its youth program in which 140 young people from Dudley and Lincoln, Massachusetts, come together in the summer to farm on lots owned by The Food Project.[3] Most of the farms' produce is sold at farmers' markets, including on Tuesdays and Thursdays on Dudley Common. In December 2009, The Food Project and DSNI agreed that the unused greenhouse would be managed by the Food Project staff to enhance the provision of fresh food through winter farming.

In the mid-2000s, local bakeries and cafés became part of new healthy food options in Dudley. In the fall of 2005, the Haley House Bakery Café

(a)   (b)

Figure 3.4

Dudley: (a) Savin and Maywood Streets Community Garden and (b) The Food Project lot 1

opened in Dudley Square to bake, prepare, sell, and cater healthy food. The café's leftover ingredients are composted and sent to The Food Project, which offers free produce to the café. In addition, the Haley House Bakery Café chef organizes healthy eating and cooking classes for at-risk kids through a partnership with local schools and a Roxbury police officer. Through Take Back the Kitchen classes, children learn to go beyond any negative preconceived opinions they might have about certain foods and about each other. The Haley House works in partnership with the Body by Brandy gym to incorporate Brandy's kids into its programs.

Environmental revitalization in Dudley also has targeted housing issues. The Dorchester Bay Economic Development Corporation (Dorchester Bay EDC) rallied forces with three other Community Development Corporations (CDCs) that were members of the Fairmount/Indigo Line CDC Collaborative to revitalize the MTBA commuter train's Fairmount Line. The revitalization of the Fairmount Line includes civic organizing partners, including Project Right, DSNI, the Conservation Law Foundation, and ACE. As part of this project, in 2009, Dorchester Bay EDC completed the construction of fifty units of affordable rental housing and 6,260 square feet of commercial space called Dudley Village. Residents cleaned up an empty lot and built a park and playground. The complex is a model of a green affordable rental property built to Energy Star II levels and featuring a photovoltaic array (Dorchester Bay EDC 2011).

More than 650 parcels of vacant land have been cleaned up and redeveloped in Dudley's core area. Residents have worked with supporting organizations and groups to achieve socioenvironmental gains such as community gardens, farms, infrastructure, playgrounds, community centers, parks, and green housing. Poverty fell from 32.4 percent in 1989 to 27 percent in 2008, and unemployment decreased from 16.3 percent in 1990 to 10.5 percent in 2011 (Medoff and Sklar 1994; Dudley Street Neighborhood Initiative 2008). Unlike other low-income and minority neighborhoods in Boston, such as Mattapan and East Boston, Dudley has provided residents with environmental and social services and resources that have much improved their quality of life. Even so, socioeconomic indicators can be improved, and joblessness and poverty are still high in Dudley. Today, residents, Community Development Corporations, and the city of Boston center many of their efforts on business development in the area, especially with the current construction of the Dudley Municipal Building on the site of the historic Ferdinand building in the center of Dudley Square. A summary of Dudley's transformation is illustrated in table 3.1.

Table 3.1
A summary of neighborhood changes and challenges in the Dudley neighborhood of Boston

Physical conditions	*Baseline situation*: 1,300 vacant lots in the mid-1980s due to arson and abandonment *Changes*: 650 lots cleaned and redeveloped and extensive brownfield redevelopment
Environmental health	*Initial conditions*: 54 hazardous waste sites with lead, chromium, mercury, asbestos, and petroleum constituents; lead poisoning and lead contamination; groundwater contamination *Changes*: Three successful lawsuits against businesses involved in illegal disposal of chemicals (trash dumping and chemical storage on four properties, illegally operating a trash-transfer station, and dumping noxious liquids into a storm drain) Closing of two illegal trash-transfer stations and conversion of one into a sustainable and safe business Projects to prevent lead poisoning (education campaigns, lead testing, lead-abatement strategies, compost distribution, crop selection, and raised beds)
Access to environmental goods	*Initial conditions*: Heart diseases as second leading cause of death for Roxbury residents and obesity as a primary health issue (as of 2001) *Changes*: Enhancement of healthy food options (3 urban farms, a greenhouse, a demonstration and education farm, 16 community gardens, a farmers' market, locally owned groceries, a community orchard, and a healthy bakery and café) New parks and green spaces and rehabilitation of abandoned ones (Dudley Town Common, Dennis Street Park, Mary Hannon Park) 3 new schoolyards and outdoor classrooms at the Dearborn, Emerson, and Mason schools Construction of 7 new tot lots and playgrounds (Eustis Street playground, Winthrop Playground) Green projects by neighborhood associations (rain barrels, composting, raised beds)
Healthy housing	*Changes*: Green housing (Dudley Village's 50 units of affordable green rental housing with a community park and playground) Weatherizing projects

Socioeconomic aspects	*Changes*: New community centers and multipurpose facilities (Kroc Center, Vine Street, Orchard Gardens, Grove Hall Community Center) Community gym (Body by Brandy) Poverty decrease from 32.4% in 1989 to 27% in 2008 Unemployment decrease from 16.3% in 1990 to 10.5% in 2011 Crime rates drop (-10% in Roxbury as a whole between 2004 and 2008)
Remaining challenges	High poverty rates 27% in 2008 against 20% in Boston as a whole 2009 median household income $33,300 against $56,000 in Boston as a whole Employment rate double the rate of Boston as a whole 25% of housing is owner-occupied compared to 32% in Boston as a whole 3,318 incidents of violent crime in the Roxbury neighborhood as a whole in 2007 (the second highest in Boston) 34% of residents with less than a high school degree against 20% in Boston as a whole 2006 hospitalization rate for children under 5 59% higher than Boston rate

Sources: City of Boston, Boston Public Health Commission, Boston Police Department, city-data.com, Dudley Street Neighborhood Initiative, Project Right, Alternatives for Community and Environment, The Food Project, Boston Natural Areas Network, Haley House Bakery Café, Shirley-Eustis House.

From Unbalanced Development to Fights for Environmental Improvements in the Casc Antic Neighborhood in Barcelona

Early Periods of Exclusion and Targeted Reinvestment

From Affluence to Decline: Barcelona under Franco   While Francisco Franco's dictatorship was ruling Spain from 1939 to 1975, Barcelona became an industrial center as new industries were established in its Les Corts, Poble Nou, and Zona Franca neighborhoods. In the 1950s, rapid urbanization was accompanied by highway construction, second-home ownership, and the search for a higher quality of life (Capel 2005). Barcelona also benefited from the development of a chemical and oil complex in the seaside city of Tarragona, which was close enough to support Barcelona's development without compromising the city's image (Marshall 2004). During this time, most of Barcelona's population increase was concentrated in the Old Town (Ciutat Vella), which middle- and upper-income residents left in search of less dense living conditions. These residential choices followed the vision of local planner Ildefons Cerdà (1815–1876), who created the extension of Barcelona called the Eixample. This new grid-planned district featured public spaces and greenery, spacious buildings, continuity of urban axes, and social diversity—all far away from the congested and insalubrious old town (Paz 2003).[4]

As higher-income and better-educated residents left the Ciutat Vella (Old Town), older residents and migrants with weak job and educational skills settled in there, particularly in the Casc Antic neighborhood. The newcomers were drawn to job opportunities and cheap rents, and they came from the agricultural parts of Spain—first Valencia, Murcia, and Aragon and then Castilla, Asturias, Andalucia, and Extremadura. Between 1950 and 1975, 7 million Spanish workers and their families migrated to Barcelona, Madrid, and Bilbao (Sánchez Lopez 1986). The percentage of working-class residents in the Casc Antic was 8 percent higher than in Barcelona as a whole in 1970 and 16 percent higher in 1980 (Sánchez Lopez 1986).

The newcomers to the Casc Antic endured hardship and precariousness. They had no formal houses to live in, had low-paid and insecure jobs at best, and suffered from Franco's policies which neglected housing, public equipment, infrastructure, green areas and public space in the city center. The Casc Antic was damp and cold in the winter, hot and smelly in the summer, and dirty at all times. By the 1970s, the public perception of the Old Town was that it was a prostitution-, drug-, and crime-ridden area (Abella 2004). Most residents avoided Ciutat Vella, which had become a shadow of its vibrant past.

The government neglected living conditions in the Casc Antic, but individual property owners and landlords also contributed to its degradation. By the time of Franco's death in 1975, the neighborhood was a "vertical shantytown" with additional floors added onto the roofs of buildings and all available spaces (including interior courtyards, terraces, and sheds) transformed into housing (Monnet 2002). Landlords also subdivided many apartments to earn additional income from newcomers, which increased overcrowding and unhealthy living conditions. In 1975, the Old Town's density had reached 833 residents per hectare (Abella 2004). In addition, landlords did not maintain their properties to ensure that tenants had decent sanitation and living space. In 1981, 13.5 percent of the buildings did not have at least one full bathroom, 85 percent did not have running water, and 50 percent were declared in need of urgent rehabilitation (Abella 2004).

Development, Contradictions, and Excesses in Barcelona from 1975 to the Post-Olympic Era   After Franco's death in 1975 and a few years of public debate within the city, the newly democratic municipality of Barcelona decided to promote the regeneration of public spaces and the construction of public equipment in decaying neighborhoods (Montaner 2004). There were five major objectives to rehabbing Barcelona's neglected neighborhoods—responding to residents' demands for improvements, promoting multifunctional spaces by adapting projects to new uses, fostering a transformational dynamic, creating new monument designs and increasing the cultural and architectural visibility of the city, and improving the image of Barcelona through marketing of its touristic, cultural, and recreational attractions (Borja 2004).

Several urban plans supported redevelopment projects and gave them coherence, structure, and legitimacy. The Plan General Metropolitano (PGM) authorized the conversion of industrial buildings and degraded infrastructure into collective equipment and public spaces, and the Planes Especiales de Reforma Interior (PERI) were comprehensive urban plans that promoted decision making with social organizations and professionals on priority projects (Borja 2004). Some PERI-based interventions, such as the one in the Raval, demolished entire blocks of buildings to create open spaces for universities, cultural centers, and museums (Arbaci and Tapada-Berteli 2012). Between 1975 and the 1992 summer Olympic games, Barcelona underwent a radical transformation from a "Catalan Manchester" to a dynamic city with a new image, physical structure, economic base, and social composition (Nello 2004).

The redevelopment of Barcelona took place in four stages. Until the first half of the 1980s, project managers applied the urban models developed in the Escola de Arquitectura de Barcelona and worked closely with local economists and infrastructure specialists (Marshall 2004). Much attention was given to social needs as 150 projects transformed the quality of public spaces and neighborhood plazas and promoted the "Barcelona dels Barris" (the Barcelona of neighborhoods) (Capel 2005).

The second period for urban projects began in 1986, when Barcelona was awarded the organization of the summer Olympic games for 1992. Under the leadership of Mayor Pasqual Maragall, new sports and recreational installations were constructed on the Montjuïc hill, and the waterfront and port were rebuilt for leisure activities. Unlike its practices from the early 1980s, the municipality began to negotiate directly with developers rather than build a consensus with civic and social organizations (Montaner 2004).

In the third period—the post-Olympics era—the municipality focused on improving the international markets for goods and services produced in Barcelona. Marketing campaigns promoted Barcelona as a city of trade and design fairs and congresses, a leader in higher education, and a top medical center. The idea was to promote "Barcelona more than ever" (Borja et al. 1997).

The fourth period was the end of the 1990s, when the municipality capitalized on the attention that it received from the games and launched "Barcelona New Projects"—American-style suburbs that coexisted awkwardly with more progressive transportation and sustainability projects (Montaner 2004). Diagonal Mar, for instance, is a gated park and community of eight skyscrapers, a convention center, a shopping mall, and a hotel built in an old industrial district.

All major urban developments have a dark side to them, and many produce social and environmental effects that are adverse for some groups and some neighborhoods—and Barcelona is no exception. A series of controversial projects that are comparable to urban-renewal projects in the United States have raised doubts about the capacity of Barcelona to remain a modern, just, and welcoming city. In Diagonal Mar, the gated private park contrasts sharply with the city's public squares, small streets, and tight urban fabric. In many cases, the PERI's urban plans have been accused of neglecting the social and economic structure of the areas being transformed (Casademunt, Alfama, Coll-Planas, Cruz, and Martí 2007; Borja 2004; Calavita and Ferrer 2004; Capel 2005; Delgado 2007; Lahuerta 2005; Leiva-Vojkovic, Miró, and Urbano 2007; Mas and Verger 2004; Montaner 2004; Unió Temporal d'Escribes 2004; Taller contra la Violencia Inmobiliaria y Urbanística 2006).

Architecture in Barcelona has overpowered urbanism, and formal urbanism and private developers have imposed their views on the city's space and its social uses (Borja 2004). Barcelona is now a large metropolis with services and shops mostly for transient populations (Montaner 2004). It has attracted millions of tourists who enjoy its cultural and recreational offerings, beaches, and tapas bars, but livability for the residents of its traditional neighborhoods has decreased. Today, urbanism is not based on the existing urban fabric but on a collage of successive, opaque actions on a territory. For critics, there is no homogeneity, coherence, or territorial balance in the city (Montaner 2004).

Acute Social and Environmental Effects of Urban Transformations

Demolition, Resident Harassment, and Poor Living Conditions   During the redevelopment of Barcelona in the 1980s and 1990s, the construction of new housing in the Casc Antic was mirrored by the destruction of working-class housing from the nineteenth and twentieth centuries, much of it with historical significance and inhabited by many low-income families. Although many architects and developers attempted to improve the quality of the housing stock, promote new business areas, and create new public spaces in Ciutat Vella, their projects destroyed old buildings indiscriminately with the support of local politicians and technicians (Capel 2005; Delgado 2007). Such projects affected the identity of the neighborhood, the construction of the historic memory, and the preservation of traditional landscapes and social fabric (Capel 2005).

The renovation of old, degraded, and semiabandoned buildings and the construction of high-end apartments made part of the neighborhood unaffordable for its former residents. This was especially true in a small area of Casc Antic known as La Ribera, in which gentrification spread rapidly. By the early 2000s in some streets of La Ribera, the purchase price of a home was twenty times the mean annual income per capita (Tello 2004). Even though the city of Barcelona built public housing in the Casc Antic—after being pushed to do so by community associations such as the Associació de Veïns—many older and vulnerable residents had to leave a neighborhood that they felt had become too expensive. The number of new units available did not equal the number of former units, which revealed that the urban projects promoted by the city of Barcelona did not address the housing needs of residents who were confronted with high mortgages or rents. The process of regenerating the area by destroying entire blocks of old buildings produced landscapes reminiscent of Barcelona during the civil war (1936 to 1939). Several streets also had empty lots full of the garbage and debris that contractors left behind.

Furthermore, the remodeling of the Casc Antic brought with it practices that residents considered excessively aggressive. First, the municipality passed measures that maximized the use of space in existing buildings by authorizing more floors per building and smaller apartment sizes. Second, tenants were harassed through a practice known as *mobbing* (Unió Temporal d'Escribes 2004). Mobbing involved threatening tenants to force them to leave without claiming expropriation rights; ending in-person rent collection and withholding new address information when landlords changed; actively or passively degrading apartments, buildings, and the neighborhood; offering monetary compensation to tenants to leave an apartment; and reconstructing buildings with the goal of increasing their value. According to one resident, emptying the buildings of residents was called "eliminating the insects" (Gottardi 2007). Buildings were then left abandoned, unsafe, and infested with rats.

Despite evictions, noise, dust, and debris, many residents fought to stay in the neighborhood. By the end of the 1990s, two main areas of the Casc Antic—Sant Pere and Santa Caterina—were still inhabited by low-income, mostly elderly residents (30 percent of the residents were over sixty years old) and newcomers from Latin America, Africa, and Pakistan. Those residents resisted mobbing, were rehoused in the few public housing buildings, or accepted substandard and unsanitary conditions in exchange for an affordable rent. Up to the mid-1990s, a high level of illicit activities still occurred in Sant Pere and Santa Caterina, which contributed to the lasting negative image of the Casc Antic and negative opinions of immigrants.

Severe Environmental Health Consequences for Residents   In the 1990s, many buildings in the Casc Antic were not structurally safe and required extensive repairs and maintenance (figure 3.5). Only 58 percent of the neighborhood's buildings were considered adequately maintained in 2001 against 80 percent in other Barcelona neighborhoods. In 2001, 175 buildings were in bad shape, including thirty-five in ruins (Martín 2007). Furthermore, in the mid-1990s, residents lacked core infrastructure, and many buildings received water from rooftop water deposits with a capacity of 100 to 500 liters—the building's entire water allowance for the day (Abella 2004). Many water deposits leaked, creating structural damage to buildings and substandard sanitation conditions (Martín 2007).

Other environmental health issues affected residents' quality of life and living conditions. Both visitors and residents dumped solid and liquid waste in abandoned lots, creating filth and poor hygiene conditions. Waste collection and management was substandard because the small streets of the Casc Antic were difficult for trash collectors to access, which left an impression of neglect. Families had no place to stroll or relax. Compared with other neighborhoods, Casc Antic (particularly in its Santa Caterina and Sant Pere sections) suffered from extensive delays in the creation of green areas, which exasperated residents and the Casc Antic Neighbors' Association (Ajuntament de Barcelona 2005).

The health conditions of residents in the Casc Antic were inferior to those of Barcelona as a whole and other neighborhoods. In 2002, Ciutat

Figure 3.5

Casc Antic: Empty lots and debris in the 1990s
Source: Veïns en Defensa de la Barcelona Vella

Vella was the district with the lowest life expectancy in the city and the highest infant mortality (five times higher than in the rest of Barcelona). High drug consumption was also a cause of health issues in the Casc Antic, whose rate of death by overdoses was higher than that in other parts of the city and other parts of the old town (11.5 drug deaths per 1,000 inhabitants for individuals between sixteen and forty-nine years old). Due to the high number of fragile immigrants and elderly residents, health clinics were in high demand in the neighborhood (Martín 2007) and often overcrowded with patients suffering from respiratory disease or complaining of air, noise, and ground contamination.

In the Casc Antic, youth had fewer opportunities to play sports and engage in other types of physical activity than youth did in other districts. The neighborhood's urban structure and the poor state of its existing centers led neighbors to complain. Children played in busy streets or small plazas. Many of their recreational activities occurred in crumbling buildings, streets full of rubbish, or construction zones—all temporary areas without the sense of stability and comfort that children needed (Martín 2007).

In sum, degradation, unequal revitalization, and marginalization in the Casc Antic unraveled in three stages. First, decades of neglect, upper-income flight, and the abandonment of the neighborhood by city officials took place from the first part of the twentieth century until the end of Franco's dictatorship in 1975. Second, public reinvestment, unequal development, forced expropriations, and voluntary abandonment of buildings and streets took place from the return of democracy until the mid-1990s. It is after this period that citizens complained about the non-selective destruction of parts of the Casc Antic, real estate speculation and social exclusion, and lack of improvement in environmental conditions.

Improved Environmental Conditions in the Casc Antic

The transition to democracy that began in Spain in 1975 paved the way for new civic associations that proposed improvements for Ciutat Vella (Capel 2005; Hache 2005). Associació de Veïns (Associations of Neighbors) had quietly formed at the end of Franco's regime and played a central role in planning projects and defending the rights of residents and social groups. In Ciutat Vella, many political and technical leaders from the Associació de Veïns del Casc Antic (Casc Antic Neighbors' Association) took key positions as urban managers and planners and started implementing the revitalization of the old town (Capel 2005). Although many resources were invested into revitalizing degraded areas and promoting tourism, this process also displaced 2,000 residents, destroyed 1,078 buildings (Mas and Verger 2004), did not resolve waste-management issues, and ignored the lack of recreational, sports, and green areas (Ajuntament de Barcelona 2005).

By the middle of the 1990s, citizens' movements were questioning the Barcelona model of urban revitalization with its excessive tourism, conversion of the city into a theme park, real estate speculation, gentrification, and deficient social and urban infrastructure (Capel 2007). At the end of the 1990s, an intense and multifaceted conflict exploded in the neighborhood, involving many Casc Antic residents, their supporters, and the city of Barcelona. In 1997, Promoción de Ciutat Vella SA (PROCIVE-SA), the public-private company in charge of the Old Town remodeling, began to relocate residents, and in 1999, it demolished some old buildings hoping to build a parking garage and high-end apartments (Alió and Jori 2010). While doing this work, city contractors left behind large amounts of rubbish that remained in the middle of the neighborhood for almost two years, which residents named Forat de la Vergonya (Hole of Shame). However, in December 2000, residents occupied the abandoned space (6,500 square meters) and transformed it into community gardens, a large plaza, playgrounds, and soccer and basketball fields. A key point in the residents' protests was that although the 1983 Planes Especiales de Reforma Interior (PERI) (Special Plan for Interior Reform) assigned a green space to this area, the municipality later changed the land use allowed for this area so that the parking lot could be built.

After a few months of sometimes violent disputes among residents, officials, and police, the Forat de la Vergonya evolved as a self-managed area and a collective reflection space. The new spaces on the Forat and in nearby buildings provided innovative social and environmental gathering points for activities such as bike-repair workshops, social documentaries, legal advice for immigrants, community clean-up, and social activities and festivals. Neighbors used these encounters to discuss and pursue their collective claims (Melo 2006). In 2002, several social and environmental demonstrations in the Forat and the whole Ciutat Vella district centered on the fight against the destruction of buildings in favor of a permanent green zone in the Forat and more green space in the Ciutat Vella.

In 2007—after years of conflict, police occupations, and the city's dialogue with only selected invited groups —the city of Barcelona finally agreed to build a permanent multiuse green space. The new space has been renamed Pou de la Figuera and includes trees, benches, fountains, playgrounds, a soccer and basketball field, and a new community center called the Equipament del Pou de la Figuera (figure 3.6). Attention was also paid to the resolution of neighborhood problems, those sharing the history of the neighborhood (Ortega 2008), dance and physical activity workshops for residents, and a new community garden called the Hortet del Forat, where volunteer participants plant and maintain a variety of Mediterranean vegetables and spices. In addition to gardening activities, volunteers organize educational workshops for local children and youth around environmental sustainability, plant ecology, composting,

Figure 3.6

Casc Antic: The reconstruction of the Forat de la Vergonya into the Pou de la Figuera recycling, and planting. Volunteers also prepare social events around the harvest from the garden, including a paella, couscous, and calçotada (a catalan dish based on local onions) festivals a few times a year. Volunteers also created a Mercat de Intercanvi (Exchange Market) to facilitate the exchange of knowledge, objects, capacities, and services (such as personal trainers and dance classes) among neighbors.

The garden is closely coupled with the work that some environmental groups do with immigrant populations. Today, Mescladis trains immigrant residents in food preparation and cooking and hires them in its two organic restaurants. As participants prepare dishes based on their Somali, Dominican, or North African culture, they also share their lives with customers and visitors. Through the workshops called Cuinar es un Joc (Cooking Is a Game), Mescladis also offers cooking classes to 250 children in the Casc Antic to create healthy eating habits and respect for diversity, solidarity, cooperation, and social commitment. Despite the economic downturn and a decrease in funding from the city of Barcelona and the Catalan government, Mescladis has sustained its activities and secured financial resources through the restaurant and the sale of food products.

In the Casc Antic, the reconstruction of the Pou de la Figuera was made possible by a 2004 Catalan law (Lei de Barris) that was seen as the result of years of resident and community organizations' demands and that brought much funding to degraded and vulnerable neighborhoods. Its funding was already guaranteed when the economic crisis hit Spain in 2007, and it invested 2.04 million euros in public space and green areas, 1.5 million euros in building rehabilitation and sanitation, 9.6 million euros in collective equipment, and 460,000 euros in waste management (Generalitat de Catalunya and Ajuntament de Barcelona 2006). The law was promoted by a new left-leaning municipality and regional government in order to improve living conditions for residents in a comprehensive way.

The city also built smaller green areas throughout Sant Pere and Santa Caterina to improve public spaces and eliminate waste accumulation and neglect in streets such as Carrer Flassaders and Carrer Forn de la Fonda. In addition, it renovated a plaza called Allada Vermell after residents complained about its lack of functionality and safety (figure 3.7). The project's proposal came from the Pla Communitari del Casc Antic (Casc Antic Community Plan, known as PICA) network of community organizations (Generalitat de Catalunya and Ajuntament de Barcelona 2006), and lobbying was done by residents and the environmental organization

(a)   (b)

Figure 3.7

Casc Antic: (a) New playground in Allada Vermell and (b) a municipal sports center

Grup Ecologista del Nucli Antic de Barcelona (GENAB). The municipality also created three community centers, including the Palau Alos youth center, which organizations such as Recursos d'Animació Intercultural (RAI), Fundació Comtal, and Fundació Adsis had long demanded.

Much improvement has been achieved in waste management. In 2005, the city of Barcelona constructed a 370,000 euro pneumatic system that moves waste from dispatch points to a collection point underground and has improved hygiene and sanitation by eliminating above-ground waste containers (Generalitat de Catalunya and Ajuntament de Barcelona 2006). During the planning phase, community groups and organizations such as Espai d'Entesa, Veïns en Defensa de la Barcelona Vella, and Arquitectos Sin Fronteras (Architects without Borders) provided technical advice to the municipality on the most appropriate points of dispatch through the Casc Antic. Last, environmental organizations such as the GENAB and the Neighbors' Association also successfully advocated for the creation of a small recycling center (Punt Verd) (Ajuntament de Barcelona 2005) and developed recycling awareness programs in the neighborhood.

Last, investments were made in structural and sanitation improvements in housing (such as the construction of direct access to main water networks) as well as green housing, mostly as a response to the demands from GENAB and community organizations. A public investment of 1.5 million euros targeted improvements in building rehabilitation in Santa Caterina and Sant Pere through the substitution of old water tanks on roofs, repairs to lead pipes and sewer systems, and construction of plumbing systems and solar panels for warm water (Generalitat de Catalunya and Ajuntament de Barcelona 2006). Local cooperatives such as Cooperativa Porfont, which emerged from the neighbors' movement in Barcelona, also funded and supervised the building of new healthy affordable housing in the Casc Antic in response to social and environmental needs in the neighborhood.

For years, the Casc Antic offered substandard options for playing sports and engaging in other kinds of physical activity. During the past decade, the Associació de Veïns del Casc Antic and other groups formed around sports, such as A. E. Cervantes - Casc Antic (AECCA), lobbied the municipality for a multipurpose sports facility, which was constructed in February 2010. This new structure has allowed neighbors and community organizations such as AECCA, Fundació Adsis, and Fundació Comtal to develop sports leagues and activities for at-risk youths. Since the economic crisis in Spain, cuts in municipal grants for local nonprofits have rendered the development and staffing of such programs uncertain and have required staff members to volunteer more of their time.

In sum, residents, their supporters, and a broad variety of networks have supported an activist-based vision for environmental and health improvements in the Casc Antic. Their mobilization came as a response to drastic environmental and social harms that resulted from city-led rehabilitation projects in the area in the 1980s and 1990s. More than in Boston, their protests were strongly resisted by public officials and police forces but eventually attracted municipal support and substantial city and state funding. As a result, residents have enjoyed enhanced urban sustainability and social equity that are in line with broader community development for the neighborhood. Today Casc Antic is the only neighborhood in Barcelona with a self-managed multipurpose community center and community garden, as well as green spaces and playgrounds that residents initiated. The proportion of affordable or public housing in the areas of San Pere and Santa Caterina is higher than any other neighborhood in the Old Town, and despite the presence of high rents in neighboring areas, residents and owners of traditional retail outlets and other commercial activities have managed to remain in those areas. Even so, gentrification, housing affordability, and the renovation of more streets and buildings are real problems. Many immigrant families still live in substandard conditions in overcrowded buildings. A summary of the improvements in the neighborhood and remaining challenges is presented in table 3.2.

Table 3.2

A summary of neighborhood changes and challenges in the Casc Antic neighborhood of Barcelona

Physical conditions	*Baseline situation*: 2,000 residents displaced and 1,078 buildings destroyed during the initial revitalization in the 1980s and 1990s; 58% of the buildings considered in an adequate state of conservation in 2001 *Changes*: Renovation of water, light, phone, gas, sewage, and public lighting infrastructure for €2.8 million €9.6 million of public investment in equipment
Environmental health	*Initial conditions*: Poor waste management, abandonment and unsafe public spaces, and lack of green space *Changes*: €460,000 of public investment in waste management (creation of a pneumatic waste system) Education campaigns about recycling practices and improved waste management Creation of a small recycling center (Punt Verd)
Access to environmental goods	*Changes*: Creation of a new community green area with recreational areas and sports ground (Pou de la Figuera) New community-based urban gardens (Hortet del Forat) Small green spaces (Carrer Flassaders, Carrer Forn de la Fonda) Renovation of the Allada Vermell plaza into a child-friendly green plaza Creation of a local network (Xarxa de Consum Solidari) with Community Shared Agriculture, ethical trade, and an exchange market A new sports center (Centro de Esportiu Municipal) and basketball leagues
Healthy housing	*Changes*: €1.5 million invested in replacing old water lines, repairing old sewer systems, installing plumbing systems in buildings, and installing solar panels
Socioeconomic aspects	*Changes*: 3 new community centers (Pou de la Figuera, Convent de Sant Agusti, Palau Alós) Training and hiring of immigrants in organic restaurant and catering projects (Mescladis)
Remaining challenges	A house purchase price 20 times the mean annual income per capita against 15 times in 1991 Gentrification Affordable fresh food Immigrant families living in substandard conditions Drug trafficking

Sources: Abella (2004); Ajuntament de Barcelona (2005, 2009, 2010); Generalitat de Catalunya and Ajuntament de Barcelona (2006); Idealista (2007); Martín (2007); Mas and Verger (2004); Tello (2004).

From Slow Decay and Disinvestment to Regeneration in the Cayo Hueso Neighborhood in Havana

Multiple Processes and Layers of Neglect

Social and Racial Segregation in pre-Castro Havana   After the Spanish-American War of 1898, Cuba declared itself independent from Spain and was occupied by the United States until it gained formal independence in 1902. In the early twentieth century, U.S. builders, architects, engineers, planners, and financial firms invested in Cuba to transform Havana into a corporate center and an elite-oriented urban metropolis. They used the city as an urban laboratory for new developments, including water mains, gas and sewage lines, rain drainage systems, electric streetlights, a garbage collection system, wide avenues, and street cars (Taylor 2009; Coyula 2009a, 2009b). This new infrastructure contributed to the exodus of wealthier residents from the congested neighborhoods of Habana Vieja and Centro Habana and toward the new Vedado and Miramar neighborhoods, which had broad streets, lush gardens and parks, elegant mansions, new social clubs, sophisticated cafés, theaters, and tourism facilities.

As in Barcelona and Boston, the massive flight of residents away from Havana's central city allowed new residents to move into Centro Habana, particularly Cayo Hueso. These new residents were mostly low-income, seasonal, and immigrant workers who were employed in construction, port installations, and tobacco factories (Díaz 2001). At that time, the very poor remained relatively invisible in Havana. Most lived in shantytowns in the outskirts of the city or were hidden in central Havana (Coyula and Hamberg 2003). They traditionally occupied *solares* and

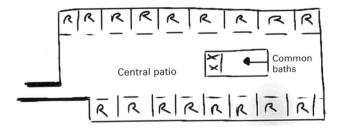

Figure 3.8

Cayo Hueso: Drawing of a typical *ciudadela* building with rooms, patio, and a common bathroom

cuarterías, which were rooming houses in baroque and neoclassical mansions that often were subdivided horizontally and vertically to build an additional floor between two official floors. They also occupied *ciudadelas*, which were one or two floors of rooms that were organized around a central courtyard and shared one common bathroom, with an entire family living in one single room (figure 3.8) (Coyula 2009a, 2009b).

In the 1940s and 1950s, sociospatial segregation in Havana was reinforced by racial segregation in housing the labor force. Afro-Cubans encountered multiple obstacles to renting or buying a home in wealthy white neighborhoods but had no problem establishing themselves in lower-class neighborhoods. They often faced landlords who claimed that apartments or rooms were already rented (De la Fuente 2000). As richer residents moved out of Cayo Hueso, large numbers of Afro-Cubans moved in. During the U.S. occupation from 1899 to 1934, American troops and white Cubans identified *solares* as spaces "for blacks" to exclude the poorest from the geography of the city and from society. It was a cultural and racial ratification of social hierarchies (De la Fuente 2000). Neighborhoods such as Cayo Hueso were also often associated with crime and promiscuity. Even when buildings or blocks were not occupied by a majority of Afro-Cubans, their residents were associated with being black (De la Fuente 2000). Afro-Cubans and women were also concentrated in the lowest-paying industrial and agricultural jobs, which meant that they often lived in the worst conditions in *solares* or *ciudadelas* (Taylor 2009).

Yet cultural traditions imbued Cayo Hueso with a strong artistic identity, which relieved the harshness of daily lives for its residents. The rumba and feeling music genres originated in the neighborhood (Gómez and Nieda 2005), and the streets and the *solares* courtyards had an artistic and lively atmosphere. Cayo Hueso was also home to Afro-Cuban *santeros* and *paleros* religious leaders (Rey, Peña, and Gutiérrez 2006) and to *yoruba* traditions, the religion brought by the first African slaves. Residents also endured the harsh and racist rules of the American occupier. American officials prohibited Afro-Cuban processions and public demonstrations by religious Afro-Cuban societies, which included what they called absurd rituals and a dangerous mixture of Roman Catholicism and evil African cults (De la Fuente 2000). Dances like the rumba were prohibited in public due to their alleged obscene and sexualized character. Afro-Cubans also had no right of entry into the numerous private clubs, casinos, and restaurants that were thriving throughout Havana (De la Fuente 2000).

*A Marginalization Exacerbated by Postrevolution Policies and Percep-
tions*   In 1959, when Fidel Castro's rebel troops entered Havana, their
message was that the revolution would support rural development and
retake control of the Cuban economy, which had been heavily influenced
by American business interests. At the end of the 1950s, Havana's population was 3.5 times higher than Cuba's next two largest cities (Santiago
and Camaguey) combined. Havana was the largest population center in
Cuba and a beacon of economic growth and cultural life, and its levels of
education, access to health care, sanitation standards, and employment
were higher than in any other part of the country (Pérez-López and Díaz-
Briquets 2000). Consequently, Castro's government focused on promoting balanced regional growth, industrialization, social equity and employment, and equal access to education and health care. New policies also
were intended to stop the migration from rural Cuba to Havana and put
in practice what Castro called in 1966 "a minimum of urbanism and a
maximum of ruralism" (Oliveras and Díaz 2007; Coyula and Hamberg
2003).

Throughout the 1960s, most public investment was directed to the
countryside and secondary cities. Between 1962 and 1972, only 15 percent of new housing developments were built in Havana even though it
contained 27 percent of Cuba's total population (Spiegel, Bonet, Garcia,
Ibarra, Tate, and Yassi 2004; Pérez-López and Díaz-Briquets 2000). Most
of Havana's new housing developments and infrastructure upgrades were
located in the shantytowns around Havana, not in the center. The absence
of a capitalist-type of market contributed to the decline of commercial
activities in Centro Habana as stores sold only basic goods or generic
merchandise (Taylor 2009) or became converted into precarious housing
(Coyula 2009a, 2009b). As a result of these readjustment policies, Havana's infrastructure and housing, especially the central areas of Habana
Vieja and Centro Habana, fell in severe decay and disrepair.

In the 1970s, the government launched a few showcase projects in an
attempt to address urban degradation (Scarpaci, Segre, and Coyula 2002).
In Cayo Hueso, Castro commissioned a project to remodel the neighborhood and promote new construction, which demolished old buildings in
a section of the neighborhood and replaced them with two twenty-story
towers (Bartolomé Bárquez 2004). According to local planners, these two
towers simply reproduced urban-renewal policies from the United States
and Europe, which systematically destroyed old, viable remnants of the
past and brought in new people. The new buildings also resulted in empty
spaces that encouraged waste dumping (Coyula and Hamberg 2003).

Under Castro's government, new forms of marginalization emerged and affected Cayo Hueso. In the 1960s, the new regime took a series of measures to address the discrimination that was faced by Afro-Cubans (De la Fuente 2000). It dismantled structures of segregation (such as private recreational spaces and schools), removed racial barriers at the workplace, and collaborated with workers' unions to ensure that the rights of Afro-Cubans were respected. However, to achieve a deracialized country, the government closed Afro-Cuban clubs because they were perceived as an obstacle to revolutionary objectives. In the 1960s, it declared that Afro-Cuban religions such as Santería were grotesque replications of primitive rites and barriers to the construction of the new Cuban man and woman (De la Fuente 2000). Some Afro-Cuban initiation ceremonies were temporarily prohibited. Even during a revolution that presented itself as egalitarian and progressive, racial prejudice existed against Afro-Cuban forms of expression (Hernández, Vásquez, Zardoya, and Mejiles 2004). Because governmental measures affected their identity, many Afro-Cubans felt excluded by the regime.

Marginalization was strongly associated with public perceptions related to crime and Afro-Cubans, which further ostracized neighborhoods such as Cayo Hueso (figure 3.9). In 1987, 31 percent of the areas officially classified as delinquency centers were in the three municipalities with most Afro-Cubans—Centro Habana, Habana Vieja, and Marianao—even though they comprised only 20 percent of Havana's population and studies demonstrated that crime rates in these areas were not above the average rates in Havana as a whole (De la Fuente 2000). Marginality in

Figure 3.9

Cayo Hueso: (a) Housing decay and (b) a typical street

the city was thus not necessarily equivalent to low access to goods or high poverty and crime (Hernández et al. 2004) but sometimes stemmed from discriminatory discourses and perceptions against certain territories and their residents.

Finally, marginality was connected to vulnerability to crises and disasters. After the economic crisis that erupted in Cuba when the Soviet Union was dismantled in 1989, an entire sector of the population became at risk or vulnerable, especially because of food insecurity (Ferriol Muruaga 1998). The economic crisis (called the Special Period) has aggravated racial inequalities and social tensions, especially in Cayo Hueso. In 1993, the U.S. dollar was legalized at twenty-six times as valuable as the Cuban peso, and is used to pay most goods and services exept for basic food, medication, and rent. This event brought about a two-tier society. Cubans who work in the tourism sector or for foreign companies are paid in convertible currency, and Cubans who work for the state receive their salary in national pesos (Coyula 2009a, 2009b). Because Afro-Cubans tend to have less access to education and highly paid jobs than white workers do, the changes in the structure of the economy have widened differences in social status and income between them and the rest of the Cuban society (De la Fuente 2000).

Territorial and Environmental Inequities in the 1980s and 1990s
In the late 1980s, Cayo Hueso was a community with substandard housing and an unhealthy environment. Buildings were crumbling and threatening to fall in on residents. Water and sanitation systems were degraded, with a strong relationship between spatial inequities and race (Coyula 2009a, 2009b). Water and sewer plants and transmission systems received minimal maintenance. Central areas of Havana were overcrowded, and multiple generations lived together. Levels of chronic heart disease, cancer, diabetes, and injuries were higher in Centro Habana than in all other municipalities (Bonet and Mas 1996). Cayo Hueso also had a high proportion of residents over age sixty-five: 13.8 percent of the population of Centro Habana was over sixty-five compared to 11 percent in all Havana and 9.1 percent in Cuba (Yassi, Mas, Bonet, Tate, Fernandez, Spiegel, and Perez 1999). Elderly residents often were not able to move through Havana to seek jobs and services and did not have access to the new market economy.

Cayo Hueso had less green and recreational space than other neighborhoods in Havana. Centro Habana had 0.22 square meters of green space per inhabitant (358 hectares), but Havana as a whole had 3.8 square meters per inhabitant (Oliveras and Díaz 2007). The neighborhood had twenty-one parks, small green spaces, recreational areas for kids, and marked public spaces, and of these, seventeen were in poor condition because of lack of maintenance, insufficient lighting, or trash dumping. Schools had little space for physical activity and often used the nearby neighborhood Parque Trillo. Children lacked access to recreational areas and playgrounds (Instituto de Planificación Física de Cuba 2002). The area was the symbol of the divisions and contrasts in the city: coastal garden-city like Havana (for visitors and privileged residents) and an inner Havana of slums (Coyula 2009a, 2009b) embodied social, territorial, racial, and cultural discrepancies in the city and revealed the role of urban territories as clear markers of inequalities.

When the Soviet Union collapsed in 1989, it triggered a socioeconomic crisis in Cuba, as the country suddenly lost its political allies, trade partners, and providers of industrialized goods and oil resources. In 1988, socialist partners accounted for 87 percent of Cuba's imports and 86.4 percent of its exports (Comité Estatal de Estadísticas 1999). In 1992, the United States tightened its embargo against Cuba, which worsened the country's economic and health situation. When the economy reached its lowest point in 1994, Cuban agriculture was producing only 55 percent of its 1990 levels (Sinclair and Thompson 2001). Especially between 1990 and 1994, the crisis had severely negative effects on health and social services, transportation, water delivery, housing, and nutrition, including shortages of meat, milk, and flour (Garfield and Santana 1997; Kirkpatrick 1997; Uriarte-Gastón 2002). From 1989 to 1993, caloric intake fell dramatically—from 3,200 calories a day to 2,100 a day (the minimum recommended caloric intake is 2,100 to 2,300 calories per day) (Food and Agriculture Organization 2001).

Reduced food supplies and increased dilapidation of the housing stock in Centro Habana and Old Havana became the most salient expressions of the Special Period. Between 1990 and 1993, housing construction came to a halt as building materials became unavailable in Cuba (Uriarte-Gastón 2002). In the 1990s, half of the 560,000 dwellings in Havana were in poor condition, 60,000 were not repairable and required demolition, and 75,000 were temporarily held up with wooden beams (Pérez-López and Díaz-Briquets 2000). Critics noted that such conditions demonstrated the failure of the Castro regime to fulfill its promise to provide decent living conditions for all (Hernández et al. 2004). Cayo Hueso was particularly hit hard by the Special Period. In the mid-1990s, 94 percent of its buildings were considered to be in "very bad" condition compared to 45 percent in Centro Habana as a whole (Instituto de Planificación Física de Cuba 2002). The precarious state of buildings left residents highly vulnerable to being injured or killed by hurricanes (Coyula 2009a, 2009b). At the beginning of the Special Period, the index of overcrowding reached seven people per room in Cayo Hueso, which led to higher rates of infectious diseases than in other parts of Havana.

Throughout the 1990s, living conditions in Cayo Hueso worsened because of failures in waste collection and maintenance of sanitation systems. Trucks collected waste only in the commercial zones, so diarrheal diseases, leptospirosis, and tuberculosis increased drastically (Instituto Nacional de Higiene, Epidemiología y Microbiología 1999). In the 1990s, more than 50 percent of residents did not have daily access to potable water (Yassi et al. 1999). Between 1993 and 1995, overall mortality in Centro Habana was 11.1 deaths per thousand population versus 7.1 per thousand in Cuba (Bonet and Mas 1996).

In the early 1990s, Havana's residents were divided between those who lived in airy, green neighborhoods and those who lived in overcrowded central areas (Espina, Moreno, Rosada, Quintana, Chailloux, and Sierra 2005). The government failed to improve conditions in the center, and traditional residential patterns linked race with poverty. During the Special Period, conditions in municipalities such as Playa were considered to be favorable and those in Centro Habana, unfavorable (Espina et al. 2005). Marginality and crime were still defined racially, which was an indicator that the goal of a deracialized society had not been achieved (De la Fuente 2000).

In sum, over a few decades of government neglect and economic crisis, residents of Cayo Hueso witnessed the slow degradation of their neighborhood. Afro-Cuban families continued to face hardships as they tried to build a place and identity for themselves in Havana and in Cuban society in general.

Autonomous Community Participation in the Rehabilitation of Cayo Hueso

Just before the fall of the Soviet Union in 1989, new decentralized community initiatives emerged in Havana. In 1987, Castro created the planning agency Grupo para el Desarollo Integral de la Capital (GDIC) (Group for the Comprehensive Development of the Capital) to advise the government on urban policy and to transition toward participatory forms of planning in Havana (Uriarte-Gastón 2002). The GDIC is a public institution that is relatively independent from Cuban central institutions. It focuses on the urban habitat as a whole to improve community life with the participation of residents (Rey 2001) and balance conservation and revitalization (Coyula 2008). In 2008, the GDIC created the Talleres de Transformación Integral del Barrio (TTIB) (Workshops for the Comprehensive Transformation of the Neighborhood) as spaces for the development of community-based solutions to problems (Rey 2001). To lead each TTIB, the GDIC assigned a multidisciplinary technical team of community members, including architects, planners, sociologists, psychologists, geographers, environmentalists, and engineers. Three workshops— Cayo Hueso, Atarés, and La Guinera—were originally created to address habitat and socioenvironmental problems in Havana's most vulnerable and deteriorated neighborhood. Over the last two and a half decades, the TTIBs have played an important role in creating and protecting an emergent civil society (Dilla and Oxhorn 2002).

In its initial years, the Cayo Hueso TTIB focused on improving housing, the local economy, education, and training and on rescuing cultural traditions (Coyula and Hamberg 2003). As it works on improving living conditions in the neighborhood, the TTIB was helped by Oxfam Canada, UNICEF and a small plant that recycled rubble into mortar blocks (Coyula and Hamberg 2003). Among other achievements, it repaired more than twelve tenement buildings in the first three years (Uriarte-Gastón 2002). One of the most successful housing improvements in Cayo Hueso was the renovation of a *ciudadela* complex on Calle Espada #411 from 1994 to 1996. Before the work was done, the building's roofs leaked, there was only one community bathroom, and plumbing was almost inexistent. Thanks to the renovation, the sixteen original single-roomed apartments were transformed into seventeen duplex apartments that each had a kitchen and bathroom and shared a common patio. Designs were chosen by each family. However, after a few years and because of the scarcity of building materials, most TTIB projects in Cayo Hueso focused on activities that required fewer material resources but still rehabilitated degraded areas (Coyula and Hamberg 2003; Díaz 2001). The TTIB turned to broader habitat and environmental improvements, such as support for high-risk groups, civic education, economic development through urban agriculture, and preservation of traditions (Rey 2001).

Throughout the neighborhood, the Cayo Hueso TTIB reclaimed empty lots by converting them into parks and recreational areas. It planted trees, developed community gardens, created community groups for environmental clean-up, designed beautification projects, developed environmental education, and implemented solid-waste recycling programs (Rey

2001). It also fixed up parks, such as the Trillo Park and the Parque de los Martires, and reforested them with the help of a European NGO and the support of the provincial government. In the health domain, the TTIB's project called Support for the Family Doctor and Nurse included the creation of twenty family medical clinics and workshops on neighborhood health issues. Its goals were to enhance working conditions for family doctors and bring medical clinics closer to residents (Díaz 2001).

One of the most innovative endeavors led by the TTIB was La Casa del Niño y de la Niña (House for the Boy and Girl), which includes a community center, an outdoor park, a playground, and recreational space that were built in 1998 on the site of a dumping ground. La Casa was the first project in Havana that was proposed by children who shared decision making with the adults working in the TTIB. The project was based on respect for children's rights (Sardiñas 2009). It has been supported by the Popular Council of Cayo Hueso, financed by UNICEF, and maintained with the help of local leaders, families, and children. In 2006, the former director of La Casa and TTIB staff members collaborated to secure the creation of the Organopónico de Alto Rendimiento, which is a high-yield urban garden that yields between 15 to 20 kilograms per square meter. Today, a Cooperativa de Créditos y Servicios (Credit and Service Cooperative, known as CCS) also offers semiprocessed food products such as jams and juices to customers next to the Organopónico, and a technical agricultural consultant (CTA) provides technical advice to residents who grow plants on patios and in backyards and sells seedlings, fertilizer, pesticides, and plants (figure 3.10).

In addition to the TTIB's work, a neighborhood movement of autonomous community organizations and activists (Dilla Alfonso, Soriano Fernández, and Flores Castro 1998, 65) began in Havana after the economic and social crisis. The neighborhood movement emerged through the horizontal networks that existed at the community level (Uriarte-Gastón 2002), and leaders acted independently and spontaneously. This neighborhood movement has been particularly active in Cayo Hueso. Local community leader Jaime and residents built Quiero a Mi Barrio (I Love My Neighborhood), an independent gym and community center on the site of two abandoned buildings. Another community leader, Cristián, developed martial arts classes in an empty lot called El Beisbolito, which he and neighbors fixed up. El Beisbolito is the only outdoor community recreational space in Cayo Hueso where children can practice sports. Save the Children is currently planning to transform it into an outdoor

Figure 3.10

Cayo Hueso: (a) A high-yield urban garden, (b) plant stalls, and (c) food stalls

gym and is also restoring playgrounds and other recreational spaces in local schools together with the support and advice of community leaders.

Furthermore, independent leaders and activists in Cayo Hueso developed environmental education and empowerment projects with children and youth. The project Moros y Cristianos was led by two Cayo Hueso residents working closely with Felix Varela, an environmental Cuban NGO. One Moros y Cristianos workshop created a Mapa Verde (Green Map) of the neighborhood. The Mapa Verde project helped "young people revitalize, reclaim, and restore their community assets" (Green Map 2007). Its ecocultural mapmaking methodology has been shared by 250 communities worldwide. In 2003 to 2004, while creating a green map for Cayo Hueso, participants developed mapping skills, gained a sense of ownership over the assets and needs of their neighborhood, and learned more about environmental protection. Children used a series of internationally recognized icons to locate environmental assets in the neighborhood, and they created their own icons that reflected their own sense of their environment. This map has galvanized community volunteers to turn empty spaces into tot lots and community centers.

Environmental improvements in Cayo Hueso have also been shaped by clean-up and rehabilitation initiatives that were led by local residents and groups. Residents clean up waste in neglected patios in their housing developments; plant bananas, vegetables, and medicinal plants; and maintain the new spaces collaboratively. Such projects are part of the eight thousand patios and *parcelas* recorded in Havana in 1999 (Caridad and Medina 2001). As part of La Casa del Niño y de la Niña, Pablo, a volunteer, together with Rosa, developed some children's brigades for environmental clean-up and protection in Cayo Hueso. Two brigades—Por un Barrio Más Limpio (For a Cleaner Neighborhood) and Arco Iris (Rainbow)—bring children together after school to do environmental clean-up and maintenance along the Malecón, in parks, in the Organopónico, and in public spaces. The third brigade, Brigada Gloop (Gloop Brigade), teaches residents how to protect existing sources of potable water in the neighborhood. Rosa, Pablo, and the children also conducted a health campaign that focused on hygiene and protection against mosquito-based epidemics, and they have lobbied the local government to increase the number of trash collectors and containers and improve street hygiene in Cayo Hueso.

Last, Cayo Hueso hosts the Callejón de Hamel, a small street that has been converted into an outdoor park with statues and benches made of recycled bathtubs, playgrounds, trees and fountains, and murals (started by the artist Salvador González in 1990) (figure 3.11). Although the Callejón was the center of "feeling" music in the 1940s, by the 1980s, it had become an unsanitary and dangerous street. Today, the street feels like an outdoor museum with murals painted with bright colors, poetry on the walls, ornate metal and concrete statues, and painted arches with vines. The project also repaired neighboring buildings, provided new playgrounds and sports facilities for nearby schools, and created artistic and recreational activities for children. It offers children a safe place to play away from heavily traveled streets. Today, the Callejón also hosts Afro-Cuban rumba concerts on Sundays, which attract visitors from all over the city and the world. Residents enjoy this vibrant community park, which is a great resource for the neighborhood. Salvador is currently working on an outdoor sculpture park and a new children's playground in an empty lot at the end of the Callejón.

Figure 3.11

Cayo Hueso: (a) El Callejón del Hamel, (b) El Beisbolito, and (c) Quiero a Mi Barrio

Since the end of the 1980s, Cayo Hueso has benefited from environmental revitalization and health improvements. These projects have been a response to earlier laissez-faire policies (of the United States and the Castro government) that allowed housing and public health problems to go unaddressed. Some local leaders worked with the Talleres de Transformación Integral del Barrio (TTIB) (Workshops for the Comprehensive Transformation of the Neighborhood), but others developed actions independently of the TTIB endeavors. Today, Cayo Hueso is the only neighborhood of Centro Habana with an urban farm, a community center such as La Casa del Niño y de la Niña, and a community gym such as Quiero a Mi Barrio. Despite the renovation of crumbling and fragile buildings through community-based diagnostics and initiatives, the neighborhood still is home to many crumbling buildings, especially the *ciudadelas*.

Renovations take place more slowly than the rate of degradation (Plan de Rehabilitación Urbana del Municipio de Centro, 2009; Coyula 2009a, 2009b), and the Cuban government does not have enough resources to invest in large-scale projects. Overcrowding is still a health and social issue. Further improvements are needed in sanitation, potable water, waste management, and water disposal. A summary of the improvements and remaining challenges is provided in table 3.3.

Discussion: Similar Patterns of Unequal Growth, Exclusion, and Rebirth

Despite their different political systems and levels of urban development, the Dudley (Boston), Casc Antic (Barcelona), and Cayo Hueso (Havana) neighborhoods reveal similar patterns of inequality, marginalization, and exclusion. Investment, growth, unequal development, and disinvestment are common in all three urban neighborhoods, even though they have not occurred at the same point in time. In the late 1980s, residents in all three neighborhoods were suffering from substandard living conditions, lack of access to green spaces and recreational facilities, and health issues resulting from inadequate sanitation, inferior waste management, and poor nutrition (in Havana and Boston). In Cayo Hueso, malnutrition was extensive, and Dudley was faced with obesity. All three neighborhoods were abandoned by public authorities, who developed a laissez-faire attitude toward illegal practices (such as trash dumping and resident harassment in Boston and Barcelona). They also had to bear negative images and stigmas about the cultural practices of residents living in the three neighborhoods and the neighborhoods' conditions in general. In other words, cities resemble each other more than countries.

After facing long-term marginalization and environmental injustices, local residents eventually joined together to reclaim and revitalize their neighborhoods' environments and health. In Dudley, inhabitants and local supporters initiated a massive clean-up and takeover of empty spaces to transform them into green spaces, community gardens, and playgrounds, while other groups and leaders focused on building technical capacity and developing new physical activity and healthy eating options. In the Casc Antic, residents spent six years in conflict with the municipality over the redevelopment of large empty lots in the neighborhood. Residents temporarily reconstructed the space as a large green area, and the municipality later rebuilt it and financed other environmental and health projects. Community attention also has focused on improvements in sanitation, building structures, waste management, additional green

Table 3.3
A summary of changes and challenges in the Cayo Hueso neighborhood of Havana

Physical conditions	*Baseline situation*: 94% of buildings considered in "very bad" state, against 45% in Centro Habana (1989); 50% of residents without daily access to potable water and living with broken waste-disposal systems (early 1990s); inadequate green and public space (0.22 square meters per inhabitant against 3.8 square meters for Havana); 81% of recreational and green spaces in less than acceptable conditions *Changes*: Construction of the Callejón de Hamel—a street converted into an outdoor park with artwork, along with playgrounds, trees and fountains, and murals; transformation of empty lots into parks and recreational areas.
Environmental health	*Initial conditions*: Chronic heart disease, cancer, and diabetes higher in Centro Habana than in all municipalities *Changes*: Solid-waste recycling programs; creation of 20 family medical clinics; kids brigades—Por un Barrio Más Limpio (For a Cleaner Neighborhood), Arco Iris, and Brigada Gloop—for neighborhood environmental clean-up and protection; more trash collectors and containers and improved street hygiene; community health campaign around hygiene and protection against mosquito-based epidemics
Access to environmental goods	*Changes*: Renovation of school playgrounds and sports facilities; construction of La Casa del Niño y de la Niña recreational space and center; Green Map project to help young people revitalize, reclaim, and restore community assets; creation of a high-yield urban garden and an agricultural advising center; rehabilitation of parks (Parque Trillo and Parque de los Martires); a community gym (Quiero a Mi Barrio) and new sports field (El Beisbolito) for martial arts classes; permaculture projects by residents; tree replanting and street beautification projects
Healthy housing	*Changes*: Renovation of more than 12 tenement buildings
Socioeconomic aspects	*Changes*: Hiring of local residents and provision of food products to residents and schools through the high-yield farm
Remaining challenges	Numerous buildings in disrepair; lack of material resources to invest in large-scale housing renovation projects, overcrowding and hygiene issues; improvements in sanitation, potable water, waste management, and water disposal; high asthma rates and infectious diseases

Sources: Bonet and Mas (1996); Coyula (2009a, 2009b); Instituto de Planificación Física de Cuba (2002); Oliveras and Díaz (2007); Rey (2001); Uriarte-Gastón (2002); Yassi et al. (1999); interviews with Plan de Rehabilitación Urbana del Municipio de Centro (2009), Miguel Coyula (2009), Roxana Mar, (2009)

space, and physical activity. In Cayo Hueso, Castro established new participatory spaces in the 1980s, and residents organized to bring about structural improvements to buildings, enhance waste management, and develop urban agriculture. Other groups coalesced around physical activities for youth and art and cultural projects. Not all changes involved only community-based efforts, but many projects were originated and supported by residents.

Despite commonalities across neighborhoods, there are differences among the three neighborhoods. First, the racial discrimination displayed by authorities, developers, and other residents was more open and pervasive in Boston than in Havana and was even less in Barcelona. In Boston, racism penetrated all spheres of life in the city and manifested itself in violent discourse and behavior. In Havana, racism was more insidious, despite the attempts of the Castro regime to eliminate racial barriers and discrimination for Afro-Cubans. In the Casc Antic, residents confronted discrimination based on their social origins and work activities rather than race, even though race has become more and more relevant in municipal and media discourses and practices. Many residents owned small traditional shops or ateliers throughout the neighborhood, which were at risk of disappearing. Older residents originally from the south of Spain were mocked for their accent and rural origins, and recently arrived residents from North Africa and Latin America are often looked at with distrust and suffer from police control. Second, Cayo Hueso residents and to a certain extent Casc Antic residents have been victims of the pervasive effects of equality policies. These policies were launched by Castro in 1959 and by the newly democratic city of Barcelona after 1975 as public officials tried to address economic inequalities and other social issues. But the policies were implemented in ways that worsened the conditions of inner-city areas such as Casc Antic and Cayo Hueso. In Boston, the authorities ignored Dudley for years.

Third, residents in Dudley confronted additional social problems in the neighborhood, such as crime and violence, which complicated their environmental revitalization work and was less relevant in Cayo Hueso and Casc Antic, where crime was limited to petty offenses and drug dealing. Finally, Dudley had more than 1,500 empty lots in the mid-1980s, and the other two neighborhoods were densely populated with few empty spaces. These different physical challenges affected community mobilization for enhanced livability and environmental quality. Today, the neighborhoods are confronted with dissimilar remaining challenges: Dudley's poverty and crime rates are still high, Cayo Hueso's core problem is the decaying state of its buildings, and Casc Antic residents face overcrowding and difficult housing conditions.

Over the past three decades, community-based revitalization in Dudley, Casc Antic, and Cayo Hueso has improved the livability and environmental quality of distressed and marginalized neighborhoods where low-income and minority residents live, learn, work, and play together. In the next chapter, I examine how activists' work reflected holistic views on neighborhood revitalization, community reconstruction and urban environmental justice, and how tactical repertoires developed over space and time.

4

Holistic Community Reconstruction and Tactical Choices

For the past two decades, activists in the Dudley, Cayo Hueso, and Casc Antic neighborhoods have spearheaded environmental and health improvements and responded to unequal development and degradation in Boston, Barcelona, and Havana. In this chapter, I examine how community activists built complementary projects that led to holistic community reconstruction. As chapter 3 shows, after these neighborhoods suffered long-term abandonment by city officials and became spatially fragmented, activists rebuilt them with environmental and health projects that complemented each other and made them more resilient, sustainable, and robust.

In this chapter, I pay particular attention to how mobilization unfolded and how activists achieved their environmental and health goals. Our understanding of neighborhood-based coalitions, especially those working for long-term environmental revitalization and not just for single issues, is limited, and comparative analyses, particularly cross-national and longitudinal comparisons of local mobilization, remain scarce, leaving many questions for researchers: What tactical methods do activists develop to enhance the environmental quality of marginalized places? Does community mobilization at the local urban level vary by political system and urbanization setting? How do internal dynamics and external contexts shape community mobilization and coalition work?

In Dudley, Cayo Hueso, and Casc Antic, activists made creative use of the local (and at times national) political context and benefited from the support of city officials and planners. They often resorted to similar tactics to achieve their objectives, including broad and flexible coalitions and bottom-to-bottom networks encompassing three forms of activism—street activism, technical activism, and funder activism.

Strengthening the Connection between Environmental Justice and Community Development

To improve the livability and environmental quality of distressed and marginalized neighborhoods like Dudley, Casc Antic, and Cayo Hueso—spaces where low-income and minority residents live, learn, work, and play together—activists have acted in a variety of intersecting domains that build on each other and reflect a natural evolution in their mind. The achievements of place-based urban environmental justice cannot be separated from community development (figure 4.1).

From Contamination to Clean-Up and Safe and Affordable Food

In each neighborhood, at an early stage of neighborhood revitalization, residents and their supporters began to clean up local land. They first fought against waste dumping and the health consequences of exposure to contaminants before turning to other environmental endeavors. By the late 1980s, Dudley residents lived in the middle of illegal trash transfers and arson, and environmental organizers confronted an acute health and human emergency. In the words of environmental organizer Alice Gomes:

I didn't have any professional background or anything, but I think when Olivia's son got bitten by something, they didn't know what it was, and he got sick. And I think that was a red flag. And I myself I had asthma back then, and my asthma seemed to be getting worse and worse.

According to organizations such as the Dudley Street Neighborhood Initiative (DSNI) and Alternatives for Community and Environment (ACE), addressing waste problems in Dudley was a multitier process that involved ending dumping, asking the city to track dumping activity, ensuring the legality of waste-management businesses, and working with business owners to create practices compatible with a safe urban environment. In 1999, ACE convinced the Boston Public Health Commission to approve new regulations on dumpster storage lots, junkyards, and recycling facilities. Similarly, in Cayo Hueso and Casc Antic, activists worked to clean up land and maintain public spaces because they believed that visible and esthetically pleasing changes would encourage other residents to help create a new neighborhood environment.

Cleaning up land eliminated contaminants and waste and helped transform vacant lots into productive spaces such as community gardens and urban farms that provide healthy and affordable food options. It moves neighborhoods from pollution to production. In Cayo Hueso,

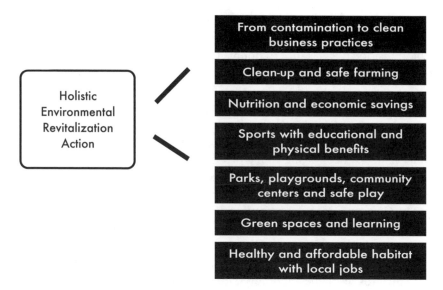

Figure 4.1

Characteristics of holistic environmental revitalization

permaculture projects and an urban farm combined land regeneration with addressing food shortages in a period of economic crisis. As Rosa from the Casa del Niño y de la Niña explained:

Things originated in permaculture. With the crisis, raising some animals and urban agriculture became permitted. Everyone started to do this. Residents were given pieces of land so that they could build a house and a garden. Soil was also given. This was done to get cheaper vegetables and to get more affordable food. We proposed [a high-yield urban garden] to the delegate from Urban Agriculture because we were concerned about the need to create an urban farm.

In Barcelona, initiatives such as Mescladis and the Xarxa de Consum Solidari contributed to the consumption of environmentally sustainable and socially just local food and provided training and job opportunities for low-income and migrant residents. In Dudley, access to fresh and affordable food was at the center of the community's activism. Environmental organizations such as the Boston Natural Areas Network (BNAN) and The Food Project provided raised beds, soil, and technical advice for community gardens that are maintained by gardeners from different ethnic groups in the neighborhood. The monetary benefits of community gardens amount to about $400 dollars per plot, which helps residents to save on the cost of their weekly groceries. In addition to technical

support, some NGOs worked to improve government policies. For instance, The Food Project partnered with the Massachusetts Department
of Transitional Assistance to double the amount of groceries for customers who use a food stamp card at weekly farmers' markets.

Outdoor Spaces and Indoor Facilities for Sports, Play, and Learning
Another core component of local activism in all three neighborhoods has
been increasing young people's access to gyms, sports fields, and community centers. Encouraging young people to play sports is a way to enhance
their physical health and, in Boston and Barcelona, decrease obesity rates.
In the Casc Antic, the mission statement of A. E. Cervantes–Casc Antic
(AECCA) states that the organization works to "protect against the diseases caused by lack of exercise, strengthen the immunological system,
improve quality of life, and raise greater awareness of the body." In Dudley, community leaders such as Brandy, an African American entrepreneur
and mentor in the neighborhood, opened fitness centers because "There
was no gym in the area and many populations of color have health issues
such as diabetes, high blood pressure, and heart problems."

Programs like Brandy 4 Kidz compensate for Boston's lower school
budgets, which have cut physical activity programs and replaced them
with standardized-test preparation. In Havana, community leaders such
as Jaime from the Quiero a Mi Barrio (I Love My Neighborhood) gym
and martial arts teacher Cristián developed sports classes for children
in spaces they cleaned up and renovated and taught children how to eat
healthy and balanced foods.

Physical activity is often combined with improving social behavior
and self-esteem. In Havana today, trainers such as Cristián, a martial arts
teacher in the Beisbolito, and Jaime, the Quiero a Mi Barrio founder, welcome children after school, promote physical activity exercises, and teach
the importance of becoming a socially responsible and educated person.
As Jaime emphasizes, his work approach is holistic:

I receive four hundred youths per day to do exercise. We help them. We get them
out of the street. We save them a little bit. They take care of their health and
their physical condition. In the educational aspect, we have workshops and discussions and seminars, based on what we can do. We look at how a sedentary
individual can end up in comparison with one who does physical activity: obesity,
high blood pressure.

The focus is on meeting individual needs, overall well-being, and the concerns of youth by taking a multisided and comprehensive approach.

For activists, developing recreational and sports opportunities also means that children are able to play safely. In Dudley, Casc Antic, and Cayo Hueso, project leaders value children's right to recreation and play as essential components of personal development and mental stability. For decades, children in dense neighborhoods did not have safe recreational opportunities, were confined at home, or played in dangerous outdoor spaces. In Barcelona, residents' efforts to rehabilitate the Forat de la Vergonya (Hole of Shame) between 2001 and 2006 ended up providing new green and sports spaces for local residents. Today a new space such as the Convent de Sant Agusti offers comprehensive benefits for local children. As Jordi Fabregas, its director, explains:

We often think about the environment as a space of quietness in the sense that we are in a gothic cloister that is . . . a space of silence in the neighborhood. It is also a playing space for kids, and it has often been converted into this. Parents get their kids from school and come here. The kids play here in the playground area while their parents are in the bar drinking a coffee.

As playgrounds and parks encourage children and youth to play outside freely, they also increase the sense of safety for families and recreate a dynamic outdoor and street life.

Some of the new tot lots and sports grounds are enclosed, as is the Kroc Center in Dudley. As Dudley activists and staff members from the Boston Department of Neighborhood Development explain, the center brings together people who did not have a place to socialize and whose outdoor sport areas are often unsafe. Unlike residents of high-income neighborhoods, many low-income neighborhoods do not have safe and welcoming public meeting places. For residents, the Kroc Center provides the conditions and amenities that a green and public space should provide, even if this space is mostly indoors.

Although residents and their supporters view physical activity and play spaces as tools to enhance local livability, their initiatives also create a dynamic balance among environmental quality, physical activity, recreation, and learning. In Dudley, projects such as the Boston Schoolyard Initiative (BSI) develop outdoor classes and schoolyards in public schools, combine educational with environmental and recreational goals, and help children create a new relationship with their place. Today schoolyards are productive environments for creative learning, playing, and socializing. Local children play during outdoor recess, which ultimately changes the relationship that they have with learning and with their neighborhood. Teachers also use outdoor classrooms during English and biology classes to enhance students' access to nature, create a more intimate relation with natural elements, and improve learning habits. Students develop writing and science skills by working on lessons and assignments in a green space. Families also use the schoolyards outside of school hours. In Havana, places such as El Beisbolito and the Quiero a Mi Barrio gym provide recreational and play opportunities for children in a caring learning environment. In such facilities, children play, practice sports, and receive training in technology and manual skill development.

From Habitat Improvement to Healthy and Affordable Housing to Economic Security

Dudley, Casc Antic, and Cayo Hueso activists advocated for the improvement of the whole habitat, particularly the rehabilitation of existing buildings. In Barcelona, groups like the Casc Antic Neighbors' Association successfully fought for upgrades in sanitation and water-delivery systems and for healthy and affordable housing for low-income families. The neighborhood group Veïns en Defensa de la Barcelona Vella connected saving historic buildings to acting for environmental sustainability because renovations consumed less energy and materials than tearing down buildings and rebuilding them from scratch. Housing cooperatives such as Cooperativa Porfont improved existing housing stock and created social housing units on public land purchased at low cost. As Oscar Martínez from Porfont explains:

As time passed, we also focused on restoration since there was so little empty land. . . . What needs to be done is to renovate old buildings, convert them fully into new ones, improve their condition—because many of the buildings out there don't even have elevators, the staircases are tiny, and the water pipes and electricity networks are very old and need to be fully renovated. The structure is also in very bad state because many are made of wood. . . . This work improves the global quality of the neighborhood because, what it achieved, is that the neighbors and their kids, who were born here, can continue to live in the neighborhood and remain integrated in it.

In the Casc Antic, a building's ground floor often is occupied by community centers, residents' associations, small sports centers, or daycare centers, which enhances residents' access to a variety of services. In Havana, the renovation of the old ciudadela building Espada 411 improved environmental safety and conditions for residents, provided green spaces in the common areas, and improved overall living conditions. A few blocks away, in the Callejón de Hamel street, artist Salvador González provided residents with materials to improve the sanitation and structural conditions of buildings. He also developed an Afro-Cuban project around public space, greening, and Afro-Cuban art and enhanced the quality of the entire habitat.

Recently, Boston and Barcelona community organizations have dedicated attention to high energy-efficiency standards and providing green spaces within public or affordable housing. Dudley Village is an example of LEED-certified (Leadership in Energy and Environmental Design) housing units built by the Dorchester Bay Development Corporation. As described by its director, Jeanne Dubois:

> For all our projects, wherever we could possibly create playgrounds or create space, we would, wherever we could do environmental upgrades. Now especially in the last ten years, there was a green Community Development Corporation initiative and we have gotten resources to use energy efficiency in our buildings. Now the new projects that we are working on here around this Quincy corridor are going to be LEED-certified green buildings. So what you are seeing in the whole CDC [Community Development Corporation] industry is a complete change into green construction—you know, green materials, solar panels, green roofs, energy efficiency.

Such projects improve the quality of the housing stock, increase residents' economic wealth, and improve the energy security of residents as weatherizing and energy-efficiency projects provide local training opportunities and green jobs. Both DSNI and ACE have advocated for energy-efficiency upgrades, building retrofits, and weatherization programs in Roxbury. For instance, the Dudley Square Organizing Project, as part of ACE, organizes residents to help them lower their energy bills. Both organizations connect just and sustainable development to broader climate-mitigation policies.

Renovated buildings and new housing provide healthier living conditions, but revitalization must keep neighborhoods affordable in order to be environmentally just. Local leaders within community organizations connect healthy and affordable habitats with environmental justice. Without the appropriate quality of buildings and without affordable rental or purchasing prices, residents will be priced out of their revitalized neighborhood. In Dudley, according to Penn Loh, former executive director of ACE, community organizations work holistically to address inequalities in the city and advocate for affordable housing in newly revitalized and greener neighborhoods:

> We realized very clearly that if we improve the environment, if we actually clean up the air, get good transit, if we get safe parks and green spaces—you know, all the good environmental justice stuff—and we haven't done anything to address

housing . . . so that they can afford to stay, then we would only be exacerbating the displacement of lower-income folks. And so that would be the ultimate tragedy—that people fight to revitalize their neighborhoods and then they can't afford to stay and they end up having to move to more marginalized areas that are less expensive but don't have all the same things that they fought for.

ACE has adapted its advocacy to emphasize housing and neighborhood affordability and joined with groups such as the Right to the City Alliance.

Toward Holistic Environmental Revitalization and Neighborhood Transformation

According to local resident leaders, community organizations, and neighborhood groups, environmental justice balances environmental care, economic development, and social protection. It requires long-term fights for neighborhood sustainability, stability, and resilience and solutions that prevent future degradation. For these three neighborhoods, tackling waste and contamination in an initial stage helped residents engage supporters and build on early successes before turning to issues of food security, parks and playground development, construction of community centers, and healthy housing. Trish Settles, a former environmental organizer within DSNI, explains the natural progression and continuum from one issue to the other:

I started out with hazardous waste sites, but you can't stop there because the disproportionate burden trickles over. And as I started talking about asthma and then kids can't go to school so parents can't get to their jobs, it became an economic-development issue. And the lead poisoning in the houses became a housing issue because we want quality housing. And then it became an energy issue: you talked about building housing with quality materials and designs so that the energy cost wasn't a burden later. Definitely environmental justice. I've talked to different colleges and conferences about it: it's much more than just open spaces and pollution, especially in an urban area where it all comes together.

Another example of comprehensive thinking is the Green Map project in Cayo Hueso. The map, designed in 2003 and 2004, tied multiple elements together by including cultural sites, green spaces, urban agriculture, historic houses, and health centers.

In sum, revitalization projects in Cayo Hueso, Dudley, and Casc Antic have transformed neighborhood conditions and habitat and increased residents' quality of life. Activists have linked environment and health and brought in tangible changes that triggered snowball effects over time. They conceived of change holistically. Community leaders and their allies anchored projects in one area of environmental justice, but their initial endeavor was a stepping stone toward related environmental initiatives. Eventually, environmental justice became intertwined with community development. In that sense, enhancing the environmental quality of distressed neighborhoods is the tip of an iceberg. Activists develop environmental quality and livability projects because they are green but also because urban environmental justice cannot exist without equitable and sustainable community development. This development takes the form of multipurpose community centers, healthy and green housing, welcoming venues for healthy food and community activities, and economic opportunities and jobs based on these projects. In return, durable community development cannot exist without environmental quality and justice.

Malleable and Multifaceted Coalitions for Community Reconstruction

Residents, community workers, and nonprofit organizations have resorted to a variety of creative tactics to advance their neighborhood revitalization projects. They used a favorable political context to assemble broad, flexible, and unexpected coalitions of supporters who acted through bottom-to-bottom networks to advance their goals. In this section, I examine the patterns of tactical choices that are common to the Dudley, Cayo Hueso, and Casc Antic neighborhoods, with some nuances between cities that reflect the adaptation of tactics to local constraints (figure 4.2).

A Clever Use of Political Context

The term *political opportunity structures* traditionally refers to the context and resources that encourage or discourage collective action (Della Porta and Tarrow 2005; Tarrow 1994). These structures allow social movements to respond dynamically and strategically to a political environment that is favorable to their demands. Movements often mount actions taking into consideration changing circumstances and political process and assessing whether certain factors (including expanding political opportunities) come together in their favor (McAdam 1982). Before making a decision, activists consider opportunities for taking action and evaluate their relationships with contenders (Tilly 1978), and their actions often build on a favorable political context (Gamson 1990). Institutional policy environments tend to affect the types of engagement that activists adopt vis-à-vis decision makers. For instance, states with high policy capacity seem to facilitate collective action, with movements operating mostly within institutional channels (e.g., Sweden); In contrast, closed systems (e.g., France) lack institutional receptivity for movements,

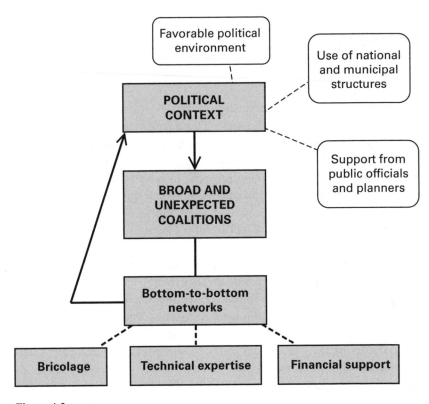

Figure 4.2
Strategic activism in the Dudley, Casc Antic, and Cayo Hueso neighborhoods of Boston, Barcelona, and Havana

and drive organizations and leaders to adopt more confrontational strategies (Kitschelt 1986). Political systems also tend to influence the initiation, forms, and outcomes of collective action (Tilly 2001). Collective action—and more direct and violent collective action—tends to arise where political repression is less likely (Tilly 1978).

Political opportunities and a favorable political context can exist as external factors explaining the existence of political action and the rise of social movements. However, in Cayo Hueso, Casc Antic, and Dudley, activists strategically used and molded this context to their goals, especially in the early stages of project development, when gaining broad acceptance was important to allow projects to move forward. Initially, the political context created incentives for community action. There was a minimum presence of state actors in each neighborhood, and their

control over local community organizing was minimal, which prompted residents to fill this space, start autonomous clean-up and rebuilding
work, and ask at times for new institutional and political arrangements
(for instance, Dudley and the power of eminent domain). In Barcelona in
2000, when it had become obvious that city administrators and workers
had abandoned the Casc Antic neighborhood, squatters occupied empty
buildings around the Forat and initiated the greening of the space. There
was a political opening in the neighborhood, and young squatters acted.
In Dudley, early environmental organizers noted that the absence of law
and order in the neighborhood was a trigger for independent community
reaction. According to Alice Gomes,

One of the problems that we were having was the transfer of trash. And people
just coming in and dumping trash in our neighborhoods. And with no penalty. . . .
We just needed to make noise, and we were just so desperate that we all just came
together to figure out how we would be heard, our story would be heard. And
how city hall would take notice.

In Cayo Hueso, because the Cuban policy making context is more
top-down and centralized, community leaders made use not only of local
political circumstances, as in Dudley and Casc Antic, but also of favorable national political circumstances and policies to initiate their projects. In the late 1980s, community members and planners shared their
concerns with Cuban officials about Havana's social, environmental, and
economic problems. Planners also expressed their frustration with the
Soviet type of central planning and demanded social initiatives that were
based on strong community participation (Rey 2001; Ramirez 2005). Because their actions took place during the Rectification Period, a campaign
launched in 1986 to "rectify errors and negative tendencies" in a context
of fiscal problems and austerity policies (Eckstein 2003), Cuban officials
welcomed their suggestions. Staff members at the Grupo para el Desarrollo Integral de la Capital (GDIC)—the planning agency for Havana's
development—took advantage of the Cuban government's support for
autonomous community-based planning and regeneration to create the
first Talleres de Transformación Integral del Barrio (TTIB) (Workshops
for the Comprehensive Transformation of the Neighborhood) workshop
in Cayo Hueso. As one former staff member explains: "It was something
that caught our attention—that we had been given concessions here to do
participatory urban planning."

Such concessions reflected the incapacity of the government to respond
to community needs. Cuban nonprofit organizations also were granted in
1993 the right to receive international funding, which encouraged TTIB

leaders to create partnerships with international nongovernmental organizations (NGOs) for their projects. In 1993, Rosa from the Casa del Niño y de la Niña linked the community's demands for an urban farm with the government's institutionalization of urban agriculture and authorization of the *unidad básica de producción cooperativa* (UBPC) (basic unit of cooperative production), a worker-managed agricultural cooperative. She and other TTIB leaders also persuaded the government of the importance of resolving the nutritional crisis in Cuba and developing neighborhood-based food-production systems. In that sense, Cayo Hueso leaders and their supporters helped to shape and benefited from national political changes in ways that assisted them in initiating their projects.

In all three neighborhoods, activists navigated existing institutional structures and organizations to promote their initiatives effectively and to receive targeted support or authorizations in regards to land use. They learned the hierarchy in municipal offices and political institutions and maneuvered through the administration to gain support for their work. The Boston Redevelopment Authority (BRA) is officially the core planning agency in Boston, but the Boston Department of Neighborhood Development (DND) allocates some city-owned land to nonprofit organizations for the community's benefit and distributes funding to groups for community gardens and open space through the Grassroots Program. Using this ambiguity, community leaders and organizations that were established and respected in the neighborhood—such as Project Hope, The Food Project, and the Nonquit Street Association—proposed new gardens and playgrounds and worked with supportive staff in the DND to create the Virginia and Monadnock Streets Community Garden and the Garden of Hope (interview with DND 2009). In Barcelona, community organizations working on sports programs, such as A. E. Cervantes - Casc Antic (AECCA) and Fundació Adsis, often called city staff members (such as the *cap de territori*, a delegate of the municipality of Barcelona who is assigned to a particular neighborhood) to gain access to schools and sports grounds for their sports league and to receive funding for their leagues.

In Havana's Cayo Hueso neighborhood, winning the support of existing official national political organizations has been crucial because municipal political leaders are sensitive to community-led initiatives. During the cleaning and remodeling of *ciudadelas*, staff members from the TTIB workshop recruited workers from the microbrigades, which are groups of thirty-five to forty people who work on construction and social projects. The microbrigades are made up of unemployed people and workers who have been temporarily released from their places of employment. TTIB members also regularly attend meetings with the Comités de Defensa de la Revolución (CDR) (Committees for the Defense of the Revolution), which is a core political organization in Cuba. They do so to demonstrate an active presence in political organizations and to assure CDR members that their organization is not losing momentum in Cuba. Staff members from international NGOs have built close relations with the GDIC planning agency to ensure greater legitimacy for their projects and find a broker for their projects. Such ties helped alleviate mistrust among community leaders, international NGOs, the municipal government, and the Office for the Regeneration and Development of Centro Habana. Oxfam's former program director in Cuba relates her experience:

It was difficult to make the government understand how it should work. People wanted to take the resources away from the workshop. The GDIC was helpful in terms of resources and was working with foreign NGOs that were bringing materials. The problem is that the construction people from the municipality wanted the materials. There were worried about a loss of control.

In all three neighborhoods, activists who consider themselves less radical and more conciliatory developed tactical and instrumental relations with local politicians. Those politicians were local political actors in Boston and Barcelona, but national ones in Havana. Community leaders and organizations used the rules of the political game, especially the fact that residents who developed strong relationships with public officials would be able to act as a voting bloc and thereby gain official support for their interests. In Barcelona's Casc Antic, the Grup Ecologista del Nucli Antic de Barcelona (GENAB) has maintained fluid communication with high-level elected officials and reminded them regularly of their electoral promises about waste management. As a result, the GENAB and its allies successfully negotiated the creation of the Punt Verd, a local recycling center in the Casc Antic. Other residents relied on public officials to gain support for the reconstruction of the Forat into a green zone. They first met the officials in other social struggles in Barcelona and found in them allies for their endeavors. For instance, Itziar Gonzalez, the former Casc Antic city manager, was committed to fostering dialogue with residents of the neighborhood and to listen to their demands. The former community relations manager in the Old Town commented on Itziar Gonzalez:

In addition to giving classes in the School of Architecture, Itziar gave technical advice to the squatters' movement in domains such as architecture. She was a mediator in the conflict around Plaza Lesseps. . . . She has an architecture office where the two architect members of the Forat squatters work, and so she advises them.

Similarly, in the early 1990s in Havana's Cayo Hueso neighborhood, TTIB leaders met with high-level politicians to persuade them to support community projects, despite their initial reluctance, as a way to avoid defections from the regime and exiles. The GDIC staff noted that officials could harness political benefits by supporting community projects. The neighborhood was necessary ("Se hacia falta el barrio"), as they pointed out. Castro recognized that the government needed to support decentralized planning and give more power to neighborhoods. In addition, community leaders negotiated for specific aspects of project development, such as the space where the Casa del Niño y de la Niña is located. As Arsenio Garcia from the United Nations Children's Fund (UNICEF) describes:

The fact is that the ministries were using the space and did not want to lose the space as part of the Ministry. However, the space was the only possibility for us. Kids needed the space to develop their capacities and do physical activities. It was a long process of negotiation. We solicited the space, then there was a delay and waiting time, and then that we achieved something with a higher level in the Ministry.

The leaders of independent projects, such as Jaime from the Quiero a Mi Barrio gym, also negotiated with public officials for the use of abandoned buildings.

In Boston's Dudley neighborhood, local organizations built relationships with political leaders that were mutually beneficial. During Mayor Raymond Flynn's administration (1984 to 1993), relations between Dudley activists and the mayor were originally tense since his traditional constituency was Irish South Boston, but Flynn soon demonstrated his support for residents of color, as Tubal Padilla, a former community organizer, notes:

Flynn became aware that someone needed to do something of great symbolic impact which would enhance the confidence that those communities had. And DSNI was helpful. And I think that our access to Ray Flynn and his disposition to act within the municipality and the municipal government to facilitate what we were doing and support what we were doing was nothing else than a . . . need to demonstrate to black and Latino communities that someone was willing to work with us.

Later on, former Boston city councilor Chuck Turner declared that the next mayor, Thomas Menino, was also able to harness political gains from his official support for Dudley's environmental projects. Menino had Italian origins and built strong relationships with local communities to ensure his political future in a city previously dominated by Irish leaders. By the end of Menino's first term as mayor, he had overseen the redevelopment of many empty lots in Dudley and continued to support community-led redevelopment in the neighborhood.

At times, however, cultivating ties with officials has been delicate for local organizations, which ultimately represent community voices. In Dudley, ACE and DSNI have navigated between engaging with city representatives and remaining legitimate voices in the community. Some members of ACE and DSNI have direct access to high-level staff within the Department of Neighborhood Development and even to Mayor Menino. Yet, they are aware that a too close relationship with decision makers may compromise their ties within the neighborhood. In Barcelona, open conflict developed between groups of activists. According to Jorge from the Barcelona Federation of Neighbors' Associations: "Once we obtained our primary goal [that it would not be a privatized space or a hard plaza], we had to negotiate. And so with this project, groups that were less compromising tended to move away from negotiations, and they continue to fight their battles." Divisions became acute between squatters and older residents (who distrusted the community-engagement practices of the municipality and wanted to solve problems on their own) and younger activists (who took more pragmatic positions).

A neighborhood is not always a homogeneous block, and different views on community relations with the municipality are constantly debated. In the community dialogues that the municipality of Barcelona organized in 2006, the city used these disagreements to engage only with groups that were favorable to its goals. In contrast, in Cuba, official political support is so important that all activists agree on the importance of building strategic and careful relationships with public leaders. In sum, in all three neighborhoods, activists have used and shaped the external political structure, organizations, and historical context to their goals.

Broad Alliances of Organizations, Groups, and People
This external context has encouraged the growth of community mobilization and allowed neighborhood activists to develop multilevel and multifaceted coalitions, which is the main tactic used in Havana, Boston, and Barcelona. In Boston, Barcelona, and Havana, developing broad alliances and coalitions is explained by the importance of implementing concrete change, cultivating strong ties, building new relationships to consolidate projects, and further anchoring them in the community.

All movements and organizations do not have access to the same resources. Resource constraints (such as limited financial resources, education levels, status, and time) constitute barriers to participation and mobilization among historically deprived or powerless groups (Fagotto and
Fung 2006; Schneider 2007). But resourcefulness and strategic capacity
(motivation, access to salient information, heuristic facility, and the presence of privileged outsiders) can allow those groups to compensate for
a lack of resources (Ganz 2000; Tarrow 2011). Skilled and charismatic
outsiders open up new options for movement leaders and can create deliberative processes that lead to important decisions (Ganz 2000).

Coalition building is also an essential tactic for social movement activists. Groups that organize into coalitions and demonstrate broad support
increase their chances of exercising power (Gamson 1990; Tilly 1978).
Coalitions are seen as more likely to emerge when resources are abundant (Staggenborg 1986; Zald and McCarthy 1987), when the identity
of movement members is cohesive, and when the collective identity of
each member is not at risk of being weakened (Diaz Veizades and Chang
1996; McCammon and Campbell 2002). They also tend to be created
when their members feel that various conditions might prevent them from
pushing for their objectives individually (Van Dyke 2003; McCammon
and Campbell 2002).

Coalitions are entities that have complementary interests and form
around a common issue or goal. Through the construction of coalitions
between groups and beyond classes, less powerful activists can eventually
put pressure on state actors and corporations (Gould, Lewis, and Roberts
2004; Polletta 2005; Di Chiro 2008; Tarrow 2011; Staggenborg 1998;
McGurty 2000). In recent years, for instance, some coalitions called "blue
green coalitions" have formed between environmentalists and labor activists (Gould et al. 2004). Networks of activists are helpful to the coalition's
work as they create links with voluntary or community organizations
or groups (Curtis and Zurcher 1973; Klandermans 1992). At the global
level, transnational advocacy networks and the coalitions within them,
which often develop around local community concerns and receive the
support of visible actors in the north, have also contributed to the success
of movements (Porta and Rucht 2002, Keck and Sikkink, 1998, Shaw
2004, Pellow 2007, Bandy 2004,). Coalitions and transnational advocacy networks are strategies that both large and small social movements
and movement organizations build to achieve their goals, as the mobilization of Dudley, Cayo Hueso, and the Casc Antic, activists indicate.

To develop broad alliances and coalitions, activists cultivate strong ties, build new relationships to consolidate projects, and anchor them in the community. In Barcelona, older residents formed coalitions in the early 2000s with the Casc Antic Neighbors' Associations, historic preservation groups such as Veïns en Defensa de la Barcelona Vella, squatters, urban gardeners from other neighborhoods, students, lawyers, architects, publishing companies, university professors, mayors of foreign cities, and movie directors. Residents also created two entities—the Espai d'Entesa and the Colectivo del Forat de la Vergonya—that protested against the Forat de la Vergonya (Hole of Shame). A third umbrella organization, the Plataforma contra la Especulació, was a Barcelona-wide coalition that fought against real estate speculation. Such broad coalitions came to include five thousand people, especially during protests and demonstrations. New organizations (such as Xarxa de Consum Solidari and Mescladis) and individual residents have joined forces to develop a community garden and healthy food-consumption and -production projects in the neighborhood.

In Boston in the late 1980s and early 1990s, residents fighting illegal dumping and trash-transfer stations and initiating Dudley's environmental revitalization connected with a variety of groups, including community organizations (such as the Dudley Street Neighborhood Initiative and Alianza Hispana), community development corporations (such as Dorchester Bay and Nuestra Comunidad), nongovernmental organizations (such Alternatives for Community and Environment and also Boston Urban Gardeners), local religious leaders (such as Father Waldron and Sister Margaret), university laboratories (at Tufts University and Wellesley College), and professors (at the Massachusetts Institute of Technology and Harvard University). New alliances have emerged around healthy eating and fresh food and include organizations such as The Food Project, DSNI, ACE, the Boston Natural Areas Network, the Haley House Bakery Café, Body by Brandy, Project Right, Boston Children's Hospital, schools, and the Boston Collaborative for Food and Fitness. A smaller alliance has formed around open-space revitalization and park stewardship and brings together the Boston Schoolyard Initiative, the Youth Environmental Network, DSNI, The Food Project, and Earthworks. Coalitions also address broader economic issues, such as green jobs, and include ACE, DSNI, and Community-Labor United.

In Cayo Hueso, community leaders such as Rosa, Jaime, and Cristián have formed broad coalitions of local teachers, doctors, musicians, and athletes to clean up and transform degraded spaces. International coalitions and collaborations have been a strong part of community projects' success. In the late 1980s and early 1990s, staff members from the Grupo para el Desarrollo Integral de la Capital planning agency initiated partnerships with university planning departments in Latin American, European, and U.S. universities. In the United States, the department of urban studies and planning at MIT worked closely with Gina Rey and Mario Coyula from the GDIC and later with Joel Díaz from Cayo Hueso to establish multidisciplinary workshops that later became the Talleres de Transformación Integral del Barrio. After international NGOs such as Oxfam joined the GDIC and local NGOs (especially Felix Varela, the Centro Martin Luther King, and Habitat Cuba), they were able to work in Cayo Hueso.

Although the coalitions formed in the three neighborhoods have been broad and diverse, their membership has been unusual in the context of environmental struggles. Artists have joined architects and environmentalists. Universities have supported community organizers. Coalitions have fluctuated over time, representing a variety of interests but uniting around common values of solidarity, altruism, and community control. They have been flexible, with relationships and contacts growing stronger during crucial moments of neighborhood organization, such as advocacy for the Salvation Army Kroc Center of Boston gym in Dudley. As Jeanne Dubois from Dorchester Bay explains: "It doesn't mean that we always get along. It doesn't mean that we don't annoy each other. But I would say we basically have a commitment to work with each other. And a philosophy of cooperation. . . . We just want to have good neighborhood partnerships."

In Barcelona, members of the Forat coalition crystallized around the rejection of development projects led by the city, with each coalition member rejecting them for different reasons. The small shops' association wanted to protect traditional trade in the area, squatters opposed speculation, and historic preservationists fought to preserve historic landmarks. Individuals and groups shared a common goal—creating a socially and environmentally sustainable place. Joan, a long-time activist and resident, explains how many different activists rallied around the Forat: "It was like the nervous center because we would all join each other there, in a certain way. The meaning of the Pou de la Figuera would be the union of all the protesting neighbors' movements. It was the emblematic place around which we would all be centered with our own fights."

Furthermore, coalitions brought together different traditions of activism for the development of community projects and built on a variety of complementary structures and experiences. In the Casc Antic, Jorge from the Barcelona Federation of Neighbors' Associations, notes that several traditions and norms coalesced during the neighborhood struggles: the Neighbors' Associations had a hierarchical structure while the newer social movements tended to be more flexible and horizontal. Such a variety of traditions joining forces strengthened the coalition, as Laia from Recursos d'Animació Intercultural explains: "Without the pressure and efforts of the neighbors from the beginning and the support of other younger social activists, who are more involved and who have more political or ideological relationships, this would not have been possible." Similarly, in Dudley, the spontaneity and rallying capacity of individual environmental organizers was complemented by the professionalism and connections of organizations such as DSNI, ACE, Project Right, and La Alianza Hispana, which offered to share their contacts with community-development experts, lawyers, and environmental policy and planning specialists.

In Cayo Hueso, the coalitions have been surprising. International NGOs such as Ayuda Popular Noruega (Norwegian People's Aid) and Oxfam America joined with respected Cuban publication venues, such as the journal *Temas*, that had similar interests in public debate and autonomous participation to ensure the smooth implementation of their projects. During the development of community projects in Cayo Hueso, *Temas* organized open debates on issues such as community participation, local leadership, racism, and marginalization within Cuban society. *Temas* reached a varied local and international crowd. It also published articles and reports on the debates, which helped legitimatize controversial projects such as the Callejón de Hamel as well as projects about autonomous reconstruction of degraded buildings.

Bottom-to-Bottom Networks, Volunteering, and Resourcefulness

As coalitions developed in Barcelona, Boston, and Havana, their members organized on many levels through bottom-to-bottom networks. Bottom-to-bottom networks are made up of people and groups that have loose and flexible connections inside (and at times, outside) the neighborhood and who actively engage in specific tasks to advance the completion of environmental revitalization projects. These tasks require physical labor and technical expertise. Here I intentionally use the concept of bottom-to-bottom networks to contrast with the traditional emphasis on bottom-up approaches in community studies and to reflect the words of activists themselves. The members of such network have strong social capital, which has been shown to foster participation in collective neighborhood action (Manzo and Perkins 2006; Saegert, Thompson, and Warren 2001). Residents rally together based on the value they assign to their place and their interactions within it, and they organize others to defend the nature of their community (Flora and Flora 2006). If used effectively, social capital indeed can generate power in the political arena by building relationships at the local level, using value-based commitments, and realizing concrete improvements (Warren 2001). Within bottom-to-bottom networks, people also use spatial capital creatively (Chion 2009) through the informal relations that they build during activities and encounters in the neighborhood streets. Activists advance their claims by taking advantage of the spatial characteristics of the place.

In all three neighborhoods, bottom-to-bottom networks were helpful as street activists pieced together material and nonmaterial resources using bricolage and collage techniques. The term *bricolage* refers to the activity of creating something by using objects that preexist in an environment. People assemble disconnected things and often act in inventive ways (Lévi-Strauss 1962). In a collective project, participants share their ideas, energy, and tools. In the Casc Antic, activists who led the reconstruction of the Forat were creative in assembling resources. Some neighbors brought plants, others built benches and playgrounds, a mayor from Sicily sent trees, older residents put together community gardens, and young people cleaned up and maintained the space. Residents used disparate objects, natural elements, and materials to create community gardens and a park and revive an abandoned area without any major outside investments. All of this was done without any major investments and in a spontaneous way, and was justified by the need to promptly address the environmental neglect and construction debris left throughout the neighborhood. Today, during celebratory events around the new community garden, neighbors continue to organize pot-luck meals based on plants harvested in the garden.

Activists in Dudley also used collage practices to achieve the construction of parks, playgrounds, and community gardens in a context of environmental damage, empty lots, and abandonment. Trish Settles, a former environmental organizer at DSNI, describes part of the process of creating a new park in Dudley:

Whether the resources are just stack time to another community, neighborhood clean-up, or "Look, hey, we've got money to put into benches." And Dennis Street Park was one of these parks where well, "Hey, we've got a resource. How do we do it? Let's put benches here and here. Oh, we've got another resource." And maybe the resources are tiny, very piecemeal for a long time. But at some point after doing a piecemeal project for a long time, it develops enough of a community momentum around it that it really drives itself.

Kathe McKenna, the founder of the Haley House Bakery Café, notes the importance of building on the close physical proximity of entities in Dudley Square to develop her healthy food project, recruit participants, and obtain fresh produce for the bakery kitchen:

We have met other groups that have tied up our connection: Patrolman Baxter partnered early on with Deedee preparing ethnic cuisine and capturing flavors in making it healthy. We partner with Body by Brandy for Kidz. She helps with obesity. We also work with Boston Day and Evening Charter Academy and take kids in a preprofessional program for culinary arts as a career.

In Cayo Hueso, several projects (such as the Quiero a Mi Barrio gym, the Callejón de Hamel, and the Beisbolito park) came about when a creative leader pieced together materials and found neighbors to help clean a degraded space. Revitalization projects took place in the context of trying to *resolver* (resolve) and *inventar* (invent) and seeking informal help from international collaborators who sent sports equipment and machines. Salvador explains how he stitched together elements to create the Callejón de Hamel:

I gathered ink from the printers that companies were throwing into the trash, and I went to get it. A neighbor would keep the paint for me in her house in a barrel. Another one got together pieces of electric wire, and I would paint at night with a light that I built. Another one would lend a ladder to me. Another one would come with his car to give me a bit of oil so that I could dilute my paints. I would use old bathtubs to create benches.

Piecing resources and materials together stemmed from a need to start projects as quickly as possible. Environmental projects were often spontaneous and unplanned, and residents had to use their inventiveness to make progress. The tactical use of collage techniques allows activists to transform the environment of a neighborhood, which in turn makes visible the neighbors' actions, gathers increased support, and enhances their legitimacy in the city. In Dudley, staff members from La Alianza Hispana note that the construction of the Jardín de la Amistad community garden allowed the organization to demonstrate concrete improvements in the neighborhood and show physical improvements in Dudley to possible funders. In the Casc Antic, community recycling projects were a pragmatic choice for activists who were looking to become more legitimate. As Albert from the GENAB explains in regards to the NGO's work on waste management: "At the beginning, we had to demonstrate that the thing was working—that it was on a good path of dialogue and pedagogy when we explained things [to the neighbors]. And yes, [the administration] started to understand, and the truth is that we have worked well." Projects supervised by GENAB or the AECCA sports association had to be viable before their members could search for broader institutional and financial support.

The environmental and health improvements that were achieved in each neighborhood could not have taken place without the active help of volunteers who were recruited by community leaders and neighborhood workers through bottom-to-bottom networking. Volunteers have been particularly important because their work makes residents aware of the importance of clean-up and environmental protection in the neighborhood. Recruiting volunteers to participate in community gardens or park stewardship work has also been a tactic that can compensate for the lack of budgets and permanent staffs. In Dudley, organizers gathered supporters inside the neighborhood for their events and activities. As Alice Gomes notes:

It was a lot of talking with people—figuring out who's available, who is not available, what they can do, what they cannot do, how can they contribute. It was a lot of work back then. Just knocking on doors. Waking up Saturday mornings and talking to folks and pointing out their rights—that we have rights and we can make a change if we all band together.

Assembling volunteer support has been translated into relying heavily on youth residents. Some organizations and groups—such as The Food Project and the Youth Environmental Network in Boston, the community garden in the Casc Antic, and the environmental brigades in Cayo Hueso—are built with young people who help staff members and community leaders run activities such as farm and park maintenance. In Havana, NGOs such as Save the Children informally built networks of young friends and colleagues in Cayo Hueso to gain access to local schools to implement their projects. Using youth and children are important tactical tools for rallying new supporters and participants in environmental endeavors. Children have a fresh and spontaneous way of integrating themselves in projects and activities, and adult leaders rely on them to recruit other volunteers. Maria from the community garden in Casc Antic explains this approach:

If you make kids participate, it is a little bit like using them, but then adults can also participate in the project if you show them how from the beginning. So the kids were the ones who decorated the fence and initiated the project. And the reception of the space was much more amicable.

Although community leaders and staff members were inventive in the ways that they stitched together resources to implement projects, they also used their networks to recruit protesters during demonstrations, sit-ins, and picketing. They aimed at rallying a maximum number of participants to criticize the actions of public agencies and private corporations and attract media attention. In Barcelona's Casc Antic, squatters, neighbors, students, and artists organized sit-ins and other demonstrations from 2002 to 2006 in the Forat and asked for an "End to the violence against the neighborhood. Put a park in the Forat de la Vergonya." They also chained themselves to the municipal building during plenary sessions to protest the speculative maneuvers of Promoción de Ciutat Vella SA (PROCIVE-SA), the company in charge of redevelopment in the Ciutat Vella. Last, press releases and manifestos were signed by many local groups and associations inside and outside the neighborhood. Activists in the Casc Antic were more radical than in Dudley and Cayo Hueso because the political context and existing traditions of activism in Barcelona allowed them to be more visible and direct in their protests.

In the early years of neighborhood reconstruction in Boston's Dudley neighborhood, residents protested against illegal trash dumping. They felt that protests were the only way to eject offenders from Dudley, as Jose Barros suggests: "It's trying to send a message. For example, when we picketed and when we went in front of this business, we decided to go around 2:00, between 1:00 and 2:00—about the time the trucks were coming back. So we blocked the gates so they couldn't enter." People also occupied public spaces in Boston, such as Mary Hannon Park, in mass numbers to reclaim the abandoned space and eliminate unwanted drug behaviors.

In Havana's Cayo Hueso, community leaders such as Rosa from the Casa del Niño y de la Niña occupied spaces that they planned to transform. Rosa joined with neighborhood children and families to occupy an empty lot and abandoned building. She also wrote a letter to the Ministry of Commerce, which owned the space, that was signed by more than five hundred children to convince the government that the neighborhood needed a new recreational area. However, Cayo Hueso leaders had to be more invisible and discreet in their contestation than in Barcelona or Boston because of the autocratic Cuba regime and the risks they might faced in the case of open protests.

Network Technical Expertise, Funding Resources, and Advocacy

Bottom-to-bottom networking has allowed coalition members who work at the street level to bring projects to the neighborhood through bricolage and collage. Activists also used networking to access technical and knowledge-based experts, who eventually became activist experts themselves. From 2007 to 2009 in Barcelona's Casc Antic, GENAB offered technical advice on sustainable techniques for urban agriculture to the organizers of the new community garden. From 2001 to 2006, the historic preservation group Veïns en Defensa de la Barcelona Vella, which has an office in the Ciutat Vella, offered legal assistance to residents who were victims of neighborhood degradation. The Barcelona-wide organization Architects without Borders has an office in the neighborhood, and its members included neighborhood squatters and friends of squatters. The group helped plan the redesign of the Forat into a permanent green zone in 2006 and 2007. Documentaries by local filmmaker Chema Falconetti became tactical tools in universities and public forums and supported residents' opposition to the controversial practices of property mobbing (pressuring apartment dwellers to vacate their apartments) and evictions. Other professionals supported the neighbors' desire to create a green zone by threatening municipal authorities with legal action. According to architect Hubertus Poppinhaus:

I filed a few denunciations in Brussels so that the municipality really implemented what they had applied for in Brussels. . . . We can notice an improvement of the municipality in that sense. . . . What they wanted to do is take the money out of the environmental fund and then, on the other hand, asphalt everything and install subterranean parking.

In Boston's Dudley, community organizations and citywide groups have provided technical environmental expertise for over two decades. In the early 1990s, DSNI and ACE, who were well integrated into the neighborhoods, asked the Massachusetts Department of Environmental Protection to declare areas as hazardous sites and obtained funding to clean up those sites. Nonprofit organizations might not have the funding, capacity, or authority to clean up sites, but they can expose the responsibility of entities such as the U.S. Environmental Protection Administration to repair discriminatory environmental practices. Organizations such as The Food Project and the Boston Natural Areas Network also offered technical support to gardeners by testing the soil and providing them with compost. Across Dudley, organizations built on close spatial ties and partnered with each other on technical aspects. As Mike Kozu from Project Right points out:

We have done a lot of stuff with ACE just in terms of environmental justice and with The Food Project as well. The Food Project was helpful especially in the farmers' market stuff that we are working on and some of the community gardens. We do recruitment for their programs as well.

Some organizations also provided planning facilitators and organized design sessions to create a unified community vision around new projects. DSNI supervised the planning phase of the Kroc Center and the construction of green housing at Dudley Village. At times, technical experts came from outside Dudley. In the early 1990s, public health studies conducted by Wellesley College and the Harvard School of Public Health helped ACE and The Food Project monitor air and soil quality and advocate for the closure of toxic facilities and the clean-up of waste sites, thereby influencing official decision making. Universities also provided training skills for garden design and landscaping. The researchers from those universities had close ties in the neighborhood through friends and colleagues already active in Dudley.

In Havana's Cayo Hueso, a series of individuals and organizations have offered technical help to the TTIB workshop members and to autonomous leaders in the neighborhood. In 1998 and 1999, the Cayo Hueso TTIB collaborated with the Instituto Nacional de Higiene, Epidemiologia y Microbiología (INHEM) (National Institute of Hygiene, Epidemiology, and Microbiology) and with Canadian universities on the Ecosystem Health project to evaluate the health and environmental aspects of the TTIB work and to document its benefits to Havana officials (interview with Annalee Yassi and Jerry Spiegel 2009). Technical expertise also helped community leaders find international partners and funders for their projects. Cuban nonprofit organizations such as the Centro de Intercambio y Referencia Iniciativa Comunitaria (CIERIC) (Information and Referral Center for Community Initiatives) and the Fundación Antonio Nuñez Jiménez de la Naturaleza y el Hombre (FANJ) (Antonio Nuñez Jiménez Foundation for Nature and Humanity) organized training workshops on grant writing, community participation, and leadership development to help local leaders develop grant proposals. As Rosa from the Casa del Niño y de la Niña explains: "To write collaboration projects is very hard. Sometimes it's forty pages with a lot of criteria. The CIERIC and the FANJ have offered classes where I went. They helped a lot on how to write based on what the other side is expecting." Last, some organizations (such as Havana's Centro Martin Luther King) engaged residents in community planning and participatory budgeting so that they could take greater ownership of renovation projects in housing buildings.

Bricolage, resourcefulness, and expertise have allowed activists to demonstrate to potential funders that project execution is based on deep roots in the community, on diverse forms and sources of civic engagement, and on technical capacity. The transformation of each place has strengthened the visibility of neighbors' action and helped them to gather more support. Building a strong fundraising capacity has been particularly important in Dudley and Cayo Hueso because of the role played by foundations and NGOs in financing community projects. In Dudley, funding for neighborhood revitalization has come from municipal, state, and federal sources (such as the Community Development Block Grants program, Department of Neighborhood Development, Boston Public Health Commission, Massachusetts Highway Department, U.S. Department of Justice, and U.S. Centers for Disease Control and Prevention), but also from local foundations (such as the Boston Foundation, Riley Foundation, and Barr Foundation), national nonprofits (such as KaBOOM! and the Salvation Army), research centers (Urban Ecology Institute), hospitals, banks (Boston Community Loan Funds), and individual donors. However, in Barcelona, the situation is different. Many activists, such as local squatters and other residents who want to remain autonomous from the state, have been vehement about not seeking and accepting municipal funding (this is not the case of more technical organizations, such as the GENAB environmental NGO or the AECCA sports association, whose funding depends on municipal sources), and the role of foundations in community revitalization is much more limited than in the United States.

In Havana's Cayo Hueso, community activists assembled a diversity of often unexpected funding sources. Funders have not been as reliable over the years as the funders in Boston but have been crucial for the development of revitalization projects. They have included NGOs (such as Oxfam Canada and Save the Children), UNICEF, cities in Spain, students and researchers from Europe and the United States, athletes from Latin America, municipal and provincial government agencies, and local residents. Save the Children also added budget categories to improve the Casa del Niño y de la Niña, repair schools and their playgrounds, and build new gyms and sports centers in the neighborhood.

Over time, some funders have become activists themselves as they reached out to municipal and national government entities to obtain support for projects (in Dudley and Cayo Hueso) or denounce local governmental practices in the press and other public forums (in the Casc Antic). In the case of Havana, Save the Children had to demonstrate to the Cuban Ministry of Education the strong partnership that it had been building with the Casa del Niño y de la Niña in order to continue working in Cayo Hueso. In Boston's Dudley, Drew Forster from the Kroc Foundation emphasizes the advocacy work that he did to convince the city of Boston to support Dudley's bid for a Kroc Center and obtain land for it:

It's the Salvation Army coming to DSNI and working with DSNI as a partner. And Mayor Menino was aware of that, I think. So obviously, we did some of our own due diligence. It's not just we found out what he had to say and we did it. . . . And I think that a lot of the public officials along the way have caught that vision and have understood the sort of "once in a lifetime moment" that this is.

In sum, foundations and large NGOs became the voice of and advocated for the neighborhoods into which they were inserted. They did not separate their funding schemes from a broader role of supporting marginalized groups that generally do not have a say in decision making. A summary of the strategic activism and the subtle differences of each neighborhood is provided in table 4.1.

Table 4.1
Strategic activism in Dudley, Casc Antic, and Cayo Hueso

	Dudley (Boston)	Casc Antic (Barcelona)	Cayo Hueso (Havana)
Clever use of the political context by local activists	• Occupation of empty political space in neighborhood • Relationship building with municipal offices (BRA and DND) and politicians at the city level (Mayor Flynn and Mayor Menino)	• Occupation of empty political space in neighborhood • Engagement with administrative staff (Cap de Territori) and officials (Regidor) at the local district level	• Occupation of empty political space in neighborhood • Clever use of new national policies (e.g., Urban Agriculture, laws on NGO funding) and structures (brigades, CDRs) and connections with national decision makers
Coalition development	• Broad and unexpected coalitions for environmental projects, including several university departments, cafés, youth networks • Combination of different forms of activism	• Broad and unexpected coalitions for environmental projects, including squatters, historic preservation groups, artists, publishing companies • Combination of different forms of activism	• Broad and unexpected coalitions for environmental projects, including Afrocuban artists, international NGOs, athletes from Latin America, and US universities • Combination of different forms of activism

Table 4.1 (continued)

	Dudley (Boston)	Casc Antic (Barcelona)	Cayo Hueso (Havana)
Bottom-to-bottom networks	• Bricolage techniques for quick project initiation to respond to chaos and deep environmental damage • Use of volunteers, especially youth and children • Creative forms of protest and resistance (e.g., human chains on streets, occupation of parks)	• Bricolage techniques for quick project initiation and response to government neglect in the neighborhood • Use of volunteers, especially children, youth, and older residents • Radical forms of protest and resistance (e.g., squatting, chaining oneself on building)	• Bricolage techniques for quick project initiation ("resolver") to respond to the crisis and lack of governmental resources • Use of volunteers, especially youth, artists, and children • More subdued forms of protest
Technical expertise and funders	• Legal and technical assistance from environmental NGOs, public health design schools, and community development organizations • Engagement with foundations and with the municipality, at some level	• Legal and technical assistance from lawyers and social architects • Little engagement with the municipality, except for more technical organizations (GENAB environmental NGO and sports groups such as AECCA)	• Legal and technical assistance from health research centers, centers for community initiatives, international NGOs • Engagement with international NGOs
Experts and funders as advocates	• Negotiation by foundations with local government	• Filing of official claims by experts	• Negotiation by international NGOs with national governments

Discussion: From Reactive to Proactive Environmental Justice and Communitywide Reconstruction

In the Dudley, Casc Antic, and Cayo Hueso neighborhoods, environmental justice is intertwined with community development. Community development becomes a tool for advancing environmental justice and vice versa. Activists have participated in a variety of complementary domains, including clean-up, safe farming, green spaces, learning, physical activity, and education. These domains feed on each other and reflect a natural evolution and continuum in the minds of activists. They are the physical aspects of long-term urban environmental justice in places where residents live, work, and play together.

Community endeavors across neighborhoods reveal the relevance of an ecosystem-health approach to environmental revitalization. An ecosystem approach promotes positive changes in individual, family, and community circumstances largely by improving physical, economic, and social conditions (Spiegel, Bonet, Yassi, Molina, Concepcion, and Mast 2001; Yassi, Mas, Bonet, Tate, Fernandez, Spiegel, and Perez 1999). It examines the relatively narrow effects of social and physical environmental factors on health and moves on to see people as participants within their dynamic social and physical ecosystem. Such a view takes a whole system into perspective by emphasizing the importance of wellness—meeting basic needs such as food, energy, and shelter while doing it in an environmentally respectful way. An ecosystem-health perspective shows how activists in Dudley, Casc Antic, and Cayo Hueso built more sustainable, resilient, and robust communities for the long term.

For two decades, activists in the three neighborhoods have developed similar tactics for achieving their goals despite acting in different contexts of urbanization and political systems. Political-opportunity structures are the context and resources that encourage or discourage collective action (Della Porta and Tarrow 2005; Tarrow 1994), and they indicate the capacity of social movements to respond strategically to a favorable political environment (Gamson 1990). In Cayo Hueso, Casc Antic, and Dudley, political opportunities and a favorable political context help explain the rise of political action and social movements. However, in this study, I show that local activists have strategically used and adapted the external political context to meet their goals, especially early in project development when gaining broad acceptance was important. Activists organized autonomously and built support around them to respond to urgent needs that the local government was not able or willing to address.

Coalitions are entities that are created by groups with complementary interests, form around a common issue, and support large social movements (Tarrow 2011; Gould, Lewis, and Roberts 2004; Polletta 2005). But coalitions and transnational advocacy networks are strategies that are used by both large national and regional social movements and small local movements. Activists in neighborhoods of Barcelona, Boston, and

Havana built broad, malleable, multifaceted, and multitiered coalitions of supporters. Coalition members in the three areas displayed similar interests in neighborhood reconstruction and community autonomy and common values of solidarity, altruism, sharing, and defense of powerless residents. Contrary to previous research (Staggenborg 1986; Zald and McCarthy 1987), coalitions did not form when resources were abundant. Resources were scarce, and coalition members initiated their revitalization work through inventive bricolage techniques.

In Barcelona, Boston, and Havana, coalition members built on spatial and social capital and used bottom-to-bottom networks to implement environmental revitalization initiatives and to oppose existing practices in the neighborhood. Activists used existing neighborhood sociospatial dynamics and strong spatial capital. The neighborhood landscape is both a material and symbolic resource for activists. The three complementary levels of activism in Dudley, Casc Antic, and Cayo Hueso include street activists, technical expert activists, and funder activists. Street volunteers participated in demonstrations and protests and gathered material and nonmaterial resources to jumpstart projects. Bottom-to-bottom networks also allowed activists to benefit from the technical expertise of lawyers, architects, planners, and environmental health specialists who either were located in the neighborhood or had close ties to neighborhood leaders. Those experts provided technical, legal, and scientific skills that could be used in city, state, and federal negotiations.

By proving their commitment to the neighborhood and the effectiveness of their networks, activists attracted diverse funding sources that have advocated for community projects. Contrary to the findings of previous research (Hunter and Staggenborg 1986), nongovernmental organizations (such as the Kroc Foundation and the Riley Foundation in Dudley and Oxfam in Cayo Hueso) have not encouraged local community organizations to turn away from more radical claims. Instead, these foundations value the community, feel strongly connected to it, and at times became activists themselves.

Differences in political systems had a limited effect on residents' activism in the three neighborhoods. To implement their initiatives, activists creatively used existing or new political spaces and structures and obtained the support of higher officials or planners in the city. Despite living in an autocratic country such as Cuba, Cayo Hueso residents were given autonomous spaces for project planning and implementation through the TTIB workshops. The Castro regime did not block the independent initiatives of other residents who created projects on their own. After forty years of dictatorship in Spain, the transition to democracy gave birth to a vigorous civil society that has contested neighborhood conditions and undertaken autonomous projects in Barcelona, a city that has been under constant redevelopment since the 1980s. Citizens have acted on their own by occupying physical and policy spaces that were ignored by the municipality. Although Boston is part of a long-term democracy, Dudley's broken local democracy meant that the demands of Dudley residents were ignored until the mid-1980s, they had no forum for expressing their concerns to policy makers, and residents had to act independently.

In other words, political and economic contexts—whether residents lived in democratic or authoritarian political systems or in developed or less developed cities—have not profoundly affected the tactical repertoires in the three neighborhoods. Commonalities indicate that experiences of marginalization, abandonment, and exasperation are closely associated with certain tactical choices (such as building large coalitions and tight bottom-to-bottom networks) to get work done without waiting for an absent government to act. At the local level, national policies or constraints (such as living in an autocratic regime) did not matter as much to activism development because local authorities were not as present and residents and their supporters had much leeway to develop their strategies. National authorities in Cuba also were interested in physical and environmental improvements to the neighborhood and accommodated innovative arrangements.

Core neighborhood tactics were similar in Boston, Barcelona, and Havana, but activists adapted street, technical, and funder tactics to each context. Because of their varying political contexts, nuances between cities are present in transparency of resource seeking and sharing (less in Havana, medium in Boston, and high and visible in Barcelona) and in length and intensity of contestation (less in Havana, medium in Boston, and high in Barcelona). In an autocratic and centralized regime like Cuba, open protests are not possible. Instead, complaint letters were written by core members of the Cuban revolution (children), signed anonymously by a wide variety of residents, and sent by a community leader (Rosa from the Casa del Niño y de la Niña) who had strong ties to political officials. An autocratic regime like the one in Cuba does not control neighborhood revitalization work as tightly as it controls political activism that has national relevance. If neighborhoods are healthier and more livable, the Castro regime knows that it might survive longer. Finally, coalition members in the three neighborhoods had to approach fundraising differently. Cayo Hueso residents sought international funding because the government was unable to invest many resources in neighborhood revitalization. Casc Antic residents were able to obtain public resources after years of civic demands and fights. Dudley residents benefited from a combination of private and public investment, which is often the case in American cities.

Much visible transformation has been achieved in Cayo Hueso, Dudley, and Casc Antic in places that residents feel emotionally connected to. Residents of historically distressed communities are often strongly attached to their neighborhood and its history. As described in this chapter, these neighborhoods support tight networks of neighbors, friends, relatives, and activists. In the next chapter, I examine the relevance of place in neighborhood activism and the ways that local activists grapple with issues of neighborhood attachment, place reconstruction, and sense of community, and in return use their attachment and sense of community to feed into project development.

5

Place Remaking through Environmental Recovery and Revitalization

Residents, community leaders, and neighborhood workers in the Cayo Hueso, Dudley, and Casc Antic neighborhoods of Havana, Boston, and Barcelona have engaged in comprehensive efforts to address long-term environmental and health inequities. Community reconstruction moved forward in these historically distressed communities thanks to broad flexible alliances. While living or working in a neighborhood, people can develop a tight sense of community and share daily experiences with neighbors and residents. In the area of urban environmental justice little research has been conducted on the relationship between sense of place and community organization, and even less research has been done from international and comparative perspectives. What is the significance of each neighborhood for activists? How do residents and their supporters experience and interpret neighborhood degradation and disruption? How is activist engagement in projects such as neighborhood clean-up, park development, and community gardens affected by a people's sense of place and identity? In this chapter, through the stories, accounts, and observations of residents and their supporters in Dudley, Casc Antic, and Cayo Hueso, I examine how an external context of degradation and marginalization shaped people's relationship to their neighborhood, their local identity, and their local engagement.

Community engagement in socioenvironmental projects was initially inspired by place attachment and feelings of belonging, grief, and loss, but these motivators were reinforced by a sense of responsibility, a desire for personal growth, and a commitment to urban nature. As residents and their allies developed these projects, they attempted to address environmental trauma by achieving environmental recovery and rootedness, creating safe havens, and celebrating the community. Comprehensive neighborhood reconstruction created intangible benefits that addressed both physical and mental health dimensions of environmental justice. Activists

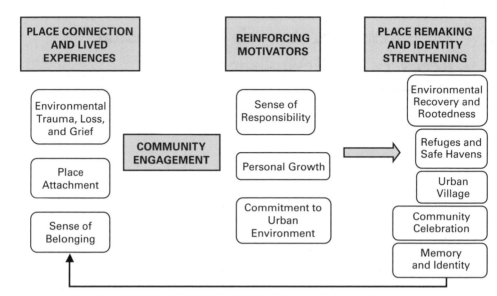

Figure 5.1

The dynamic relationships between place attachment, community engagement in environmental justice work, and place remaking

remade a place and home for marginalized residents in the city and developed a stronger collective identity. In return, a deeper connection to place and a stronger place-based identity encouraged long-term resident engagement in and for their neighborhood. In other words, place remaking was a dynamic and dialectic process (figure 5.1).

Lived Experiences in the Neighborhood

Place Attachment, Sense of Belonging, and Emotional Connection

Place attachment is the affective bond between people and places (Low and Altman 1992). It can rests both on physical features and social dimensions (Scannell and Gifford 2010). In Dudley, Cayo Hueso, and Casc Antic, community activists expressed their attachment to their place in multiple ways. They extracted from spaces a clearly identified, bounded, and meaningful place (De Certeau 1984). First, people valued the physical assets of the neighborhood and valued living in a centrally located area that was connected to other parts of the city and offered amenities (traditional shops, transportation connections, artisans, and small businesses) to its residents. Amenities and centrality create foot traffic, circulation of people and goods, and street activity. In Barcelona, several long-time community leaders mentioned that the Casc Antic is well connected to the port, the França train station, the upper town, and industrial districts. They also spoke of the many commercial and artisan activities (such as cotton makers, furniture sellers, and small printing offices) that made the neighborhood a vibrant and dynamic place. They express a strong symbolic connection to their place. In Havana, community leaders within the Talleres de Transformacion Integral del Barrio (TTIB) (Workshops for the Comprehensive Transformation of the Neighborhood) remembered that many commercial activities once thrived throughout the Cayo Hueso and that tobacco factories converted the neighborhood into a center of cigar production in the earlier decades of the twentieth century. They can even tie many of their ancestral roots to the neighborhood. In Dudley, residents felt proud to be near the center of Boston and live in the middle of landmarks such as the Shirley-Eustis House, but also felt remote from it mentally because the thriving streets of downtown Boston contrasted dramatically with the less prosperous areas in their neighborhood. Massachusetts Avenue was the physical divider between these two worlds. Residents were aware of Dudley's rich and vibrant past but also were conscious that effort was needed to continue the durable transformation of their neighborhood.

Second, residents were strongly attached to the cultural and social patrimony of their neighborhood, and this patrimony contributed to their place attachment. This was true for both older activists as well as younger ones. In Barcelona and Havana, residents felt honored to live in a part of the city with strong traditions and artistic talents. In Cayo Hueso, many leaders remembered rumba and son musicians, such as Chano Pozo, who entertained residents in old *solares* (rooming houses) in the neighborhood. Older residents talked enthusiastically about the past grandeur of their neighborhood and noted the old houses, churches, convents, and ornate building facades that they valued as community jewels. They appreciated the importance of preserving old palaces and señorial houses from the eighteenth and nineteenth centuries (in the Casc Antic), government landmarks such as the Shirley-Eustis House (in Dudley), and monuments such as the park and statue of Quintín Bandera (in Cayo Hueso), a revolutionary general who fought against Spanish colonization in the nineteenth century.

Residents, leaders, and community workers not only felt connected to the physical neighborhood but also to the ethnic and social groups that lived in it. They valued the intermixing of generations and cultures, as well as the waves of immigrants and migrants from other parts of the country who enriched their place. People appreciated the warm and informal relationships between residents who were sources of mutual help and trust. These relationships contributed to creating a strong sense of community. Here are the words from the leader of the Quiero a Mi Barrio gym in Cayo Hueso:

I love my neighborhood. I did not want to leave even though I had the opportunity to leave. These are my roots. I like the neighborhood and its form of all living together. In other countries, people live in their houses isolated. This is the case in Europe. You have a neighbor, and you don't know them.

The attachment to the neighborhood was thus also a question of atmosphere and familiarity and proximity between people. The neighborhood was not a cold and impersonal space for activists but a warm place where people felt close to one another.

The attachment that people feel toward a place is often connected to their sense of community, which includes the connection to a shared history and concerns (Perkins and Long 2002). Relationship building in Cayo Hueso, Dudley, and Casc Antic occurred daily and mostly outside the house. Life on the streets was important, and much time was spent there making connections by talking, playing, and doing activities with others. Strolling up and down the streets and talking to others was a way to spend time outdoors and a recognition of the value of socializing and being with each other. It was a simple and informal type of human interaction that people cherish. As Travis Watson from the Dudley Street Neighborhood Initiative (DSNI) explains:

I really like the sense of community. I feel like people kind of look out for each other. You see a lot of people that are helping the elderly off the bus. I feel like it's some of the little things I can kind of pick up in this community. I just like the vibe. I like just walking down the streets and kind of just talking to people. People are very real here.

Other environmental organizers shared similar stories about the liveliness and energy of the neighborhood, highlighting the fact that many people moved and still move to Dudley because of its intensity and because of the meaning of community for residents. It was and is not a neutral place in the history of Boston.

Part of the community-building experience happens when leaders and residents share life experiences with neighbors. In Havana, people lived through important life stages together and organized celebrations for birthdays, communions, and marriages. They also watched each other during illnesses. In Barcelona in the 1990s, social architects, squatters, and members of the Neighbors' Associations helped residents who suffered from mobbing (being pressured to leave their dwellings) and evictions. In Boston, neighbors offered support during health emergencies, some of them caused by illegal dumping and arson in Dudley. In 1987, several neighbors accompanied the Barros family when one of their sons developed a serious infection that they suspected was caused by an illegal trash-dumping facility on Robey Street operated by AFL Disposal Company. Many activists (and their parents) lived in their neighborhood for decades, which made them strongly connected to the place and to each other. As Joan, the owner of a natural medicine store in the Casc Antic, remembers:

Before, Manel and Angelita would come spontaneously to the house of my parents, or my parents would go to their house. Or Raquel would come down from the top floor where she was living. My mom still goes to her house every day to see her . . . because she loves her. . . . they are like family, even if they are friends. It is as if they belonged to the family.

Family members included relatives by blood and marriage but also residents with whom people have built close ties. For newcomers, arriving in a friendly community and attending community events helped them feel more at ease and comfortable with their new place. The neighborhood was a residence, a home, and a place where residents experienced the affection of others.

Environmental Trauma, Loss, and Grief

Even when place attachment and sense of community were strong , residents and their supporters also had severe impressions of an urban war in their neighborhood in terms of violence, conflict, demolition, fragmentation, and shock. Many activists told stories about a neighborhood that once felt and looked like a war zone (with crumbling buildings, waste, and rodents) and a victim of municipal violence toward residents. They share detailed images and vivid descriptions of their experiences of environmental trauma.

In Barcelona, activists remembered when their neighborhood looked like a war zone during the 1990s and early 2000s because of demolitions and the battle between residents and the municipality over the redevelopment of the Casc Antic, especially the Forat area. As an activist, Joan, recalls:

[The Forat] was a completely devastated area, and . . . the neighborhood had to swallow all of this filth and throughout the whole day endure it. . . . It was a real degradation, like a bomb attack—you know, as if a war had gone through. It was an infection, really.

The Forat de la Vergonya (Hole of Shame) was a site where the municipality allowed the destruction of buildings in 1999 and contractors left behind debris for years. For some older neighbors, many scenes were reminiscent of the Spanish civil war. Between 2001 and 2006, residents assembled materials to rebuild the area, but the police intervened several times to dismantle resident-made playgrounds and community gardens. The police also put up a wall to prevent people from accessing the space. Protests were at times violent: there were broken windows, verbal attacks, rocket launchers, and helicopters flying over the Forat to support police forces on the ground. Residents throughout the neighborhood felt like they were prisoners, physically and mentally restrained. When protesters took down the police wall, they compared their actions to those that took place in Berlin in 1989. Neighbors were arrested, which reinforced residents' fears and anger and gave the impression that this area of the Casc Antic had become uninhabitable. Immigrants, migrants from southern and northern Spain, and squatters all described the fights as guerilla warfare that media reported on and decision makers watched with apprehension.

In Boston's Dudley neighborhood, similar impressions of urban guerilla warfare were omnipresent in older residents' memories, and they originated in the arson, illegal trash-dumping practices, and crime of the 1970s and 1980s. Residents witnessed nightly arson, heard the strident sounds of sirens, and saw firemen storming through the neighborhood to extinguish fires and remove people from their burning houses. Trauma was exacerbated by high crime rates and street violence, and residents feared leaving their houses or allowing their children to play on the streets unsupervised. Tensions worsened when school busing was implemented in 1974. When students were bused to schools in other neighborhoods of Boston as part of court-ordered desegregation measures, children from marginalized neighborhoods became the target of violence. Charlotte Kahn, one of the founders of Boston Urban Gardeners explains: "There was a lot of violence. People were killed. There was tremendous tension, real turmoil. And there are still a lot of scars from that period."

Absence of law and order were also characteristic of this period and reminiscent of war scenes because the city of Boston did not investigate illegal trash-dumping practices or fine their perpetrators. It was "daily chaos," according to some environmental organizers. More recently, new immigrants to Dudley have experienced other kinds of traumatizing experiences, such as departure from their country of origin, separation from their families, and conflict or civil war, as is the case with many East African immigrants.

In Havana's Cayo Hueso, people put less emphasis on experiences of urban war, but both young and old residents described neighborhood devastation through building and infrastructure collapse since the 1980s, which forced many families to move to shelters. That said, in the early years of the Special Period (after the fall of the Soviet Union in 1989), sanitation and hydraulic infrastructures were on the brink of collapsing. Black water ran through the streets where children played. People walked on the streets rather than on sidewalks to avoid being hit by bricks or stones falling from roofs. The street of the Callejón de Hamel used to be a "very destroyed" cul-de-sac, according to its founder. People felt ashamed of their neighborhood. In the 1970s, the government ordered some demolitions in Cayo Hueso, which produced feelings of alienation among residents and which today still define part of the neighborhood landscape. Miguel Coyula, an architect in the GDIC planning agency, explains the impact of the 1970s urban-renewal project:

The problem is that the skyscrapers created anonymity. . . . I knew neighbors from Cayo Hueso whose point of reference was the street, and now they live on the eleventh floor without a balcony. Once, I arrived in one of those buildings and found a couch in a hallway outside the apartment. The sense of belonging is very great: "my house, my street, my block, and my bakery."

As new buildings took the place of old ones, residents became disoriented because their traditional use of space was changed by the demolition and reconstruction.

As a result of traumatic and alienating experiences, people in all three neighborhoods felt individual and collective grief and a sense of loss about their place. In the Casc Antic, expropriations in the 1990s felt like massive deportations to residents. When the municipality ordered demolitions, entire streets disappeared. People nostalgically named former streets and remembered traditional activities that used to thrive in Casc Antic, such as furniture builders, carpenters, ironworks shops, and watchmakers. They considered the destruction of buildings and the ensuing empty Forat area as a trigger to a loss of reference and memory. M. A. Santos, a project manager for the city of Barcelona, explains: "Yes, there has been a lot of people hurt, and so it is very complicated when you get immersed in issues of suffering, of feelings of being attacked. In the end, this makes the situation a little bit violent because the response is almost emotional." The space in which people used to live became an empty and unused hole after being a space that residents remembered as the house of their parents or grandparents.

Neighborhood disruptions also brought population changes and the loss of social capital and traditional networks, especially for older residents. Expropriations and displacement meant that people lost friends and family members. Neighbors they had seen daily for years got relocated to other neighborhoods. Residents had to move to areas where they did not know anything or anyone. People became disoriented because their physical, social, and mental markers were erased and their appropriation of space was affected. The loss was multifaceted, and people talked nostalgically about the traditional fabric and interpersonal relationships of the neighborhood. This was particularly true in the Casc Antic because municipal plans had a devastating effect on the neighborhood in a relatively short amount of time. In many cases, people had to move from the Casc Antic to the suburbs or to the Barceloneta.

Experiences of physical and mental displacement also created fears of erasure from the neighborhood and city (Manzo, Kleit, and Couch 2008). People felt that their neighborhood was not their own anymore. Nelson Merced, the first director of Alianza Hispana, a community organization in Dudley, relates this fear and connects it to the threats of encroachment from businesses and residents in the South End:

You know it was a real fear. It wasn't a fictitious fear. It was a fear that people saw. They had seen the consequences of when the city and state decide to do something. And so they—you know, I think they were highly aware of that possibility and decided that they were going to continue to take action.

In the Casc Antic, supporters of the fight to clean up the abandoned Forat site expressed their disapproval of municipal plans that they felt wanted to create "a new identity by erasing the plaza" (according to film director Chema Falconetti). This fear of erasure was felt by older residents and also by residents who arrived more recently in Dudley and Casc Antic and who expressed their fear of losing their homes and being displaced from the neighborhood. In Cayo Hueso, Afro-Cuban leaders connected their fear of erasure to the Cuban regime's attempt to assimilate them and their culture into the white socialist mentality and to eradicate their artistic and culture practice from the public eye. In official discourses and laws, the government attempted to foster equality between people, but in daily interactions and in public spaces, Afro-Cubans in Cayo Hueso felt like second-tier citizens.

As much as older and more isolated residents experienced loss and expressed grief toward their former lives, local youth also expressed disenchantment about their future. First, many children were separated from parents who fled Cuba or did not immigrate with them to Barcelona or Boston. They also were neglected within their own family, which affects their physical and mental health. Many lived and still live in low-income households, their caretakers were under- or unemployed, and self-confidence was undermined. Second, growing up in a declining neighborhood or moving to a half-destroyed place prevented them from projecting themselves positively. Several youth groups in each neighborhood said that they did not know where they belonged any more or how to use their neighborhood. In Barcelona's Casc Antic, youth felt ungrounded in their place and disconnected from their neighborhood, their city, and the image of modernity projected by the city. In Boston, community leaders often noted that young people felt disconnected from their neighborhood and city and lacked a sense of rootedness. For this reason, several funders promised not to take away any asset or point of reference from young residents when they helped to rebuild the neighborhood.

In sum, in all three neighborhoods, residents and their supporters shared stories that revealed that place attachment and positive experiences related to place were threatened and affected by experiences of environmental trauma, loss, and grief.

Feelings of Responsibility
As a result, in the Dudley, Casc Antic, and Cayo Hueso neighborhoods, deep attachments, community belonging, emotional connections, and place identity motivated residents to take action and improve their place. Community identity and place attachment have been shown to encourage participation and collective action (Cox 1982; Davis 1991; Gotham and Brumley 2002; Suttles 1968; Tilly 1974). In the three neighborhoods, attachment and identity precedes activism but also reinforce it. In Havana, Arsenio Garcia, the UNICEF coordinator who has worked closely with Rosa, the founder of the Casa del Niño y de la Niña, explained that Rosa's feelings of belonging to Cayo Hueso motivated her to build the house and the playground next to it: "Rosa did everything in what the house is now. She has another meaning for what she does. She has a strong sense of belonging, a lot of energy. But she had to overcome many difficulties." Another community leader, Jaime, called the new gym that he created Quiero a Mi Barrio (I Love my Neighborhood), suggesting that his endeavor was based on his connection to the place.

In Barcelona, a staff member of the environmental organization Grup
Ecologista del Nucli Antic de Barcelona (GENAB) explains the progression from attachment to action:

When you belong to some place, you love it, you take care of it, you spoil it, you
want it to improve, no? And put into practice, this feeling means that, without
much fuss, you focus your day-to-day work on the project that you want the
most. . . . It's not anymore "I have to do this."

Coupled with place attachment, the experience and history of a community that was going through trials and accomplishing change encouraged activists to remain organized and participate in efforts to enhance
the environmental quality and livability of their neighborhood. Residents
felt responsible for following in the footsteps of others who fought to
improve the quality of life for families. This sense of responsibility was an
additional motivator for action. As a squatter in Barcelona emphasizes:

I got involved in the neighborhood fights because it was abandoned, because of
the Forat, because I wanted to do something for the neighborhood, and because
there were already occupied houses in the Forat. . . . And because the movement
had already been going on for years and something really beautiful was being created with the people who were living there, that gives you a desire to participate
in this.

People were aware of the history of each neighborhood and were committed to playing their part and contributing to its transformation.

The testimonies of residents and their supporters reflect common tra-
jectories of the activists who participated in protests, sit-ins, demonstrations, and local mobilizations. They felt proud of being part of a successful
grassroot organization that brought them closer together and built their
collective memory of a place. Past struggles and early accomplishments
were woven into the story of the actors, pieces, and milestones that made
up a neighborhood. Activists explained how they embraced early victories
in the 1990s as signs that a different future might be possible. Successful land clean-up and community garden construction in the Casc Antic
and Dudley changed the appearance of the neighborhood and thereby
the image that residents had of it. The neighborhood became beautiful,
attractive, and welcoming—again. In Havana, the clean-up of the space
where the Beisbolito sports ground is now located or the reconstruction
of the Espada 411 *ciudadela* courtyard building produced a similar sense
of possibility. Early positive changes led people to imagine new uses for
old degraded spaces and new such as parks and recreational grounds.

The sense of duty that local activists expressed toward their neighborhood led them to desire to give back to the community. They saw

themselves as made by the neighborhood and by the relatives who raised them and created a sense of home inside and outside the house. Leaders such as Jaime and Joel Díaz from the TTIB workshop in Cayo Hueso; environmental organizers Alice Gomes, Travis Watson, and Tubal Padilla in Dudley; and long-time residents and social architects in the Casc Antic felt indebted to their place. For example, Joel Díaz was hosted by Cayo Hueso residents during his university studies and expressed gratitude toward the residents who welcomed him. In Dudley, the words of Alice Gomes are revealing: "It was my community. It was the only community I knew. I mean, I grew up there. My parents still lived there. I had great times there. So how can you not do anything? That was home for me. . . . You know, when you see something is not right, you have to address it." The moral values of those activists could be expressed as taking care of their neighborhood and preserving improvements that have reconstructed the place. They wanted new generations of residents to experience the same attention and care that benefited them.

Many residents and community workers stayed in the neighborhood, even though they recognize that they could have left. Large numbers of black middle-income residents have not fled Dudley for the suburbs of Boston. They have served as a central asset for increasing Dudley's economic stability, social cohesion, and political mobilization. Black middle-income leaders organized other residents to participate in neighborhood projects and cultivated positive attitudes and civic energy within the community. DSNI's executive director, John Barros, exemplified this commitment of younger and educated residents who returned to Dudley after their college studies. John studied at Dartmouth College and came back to Dudley to apply his new planning and political-analysis skills to the further redevelopment of his neighborhood. In Barcelona, many residents could have left for nearby affordable neighborhoods such as the Barceloneta or for cheaper suburbs, but they chose to stay. In Cuba, residents tended to be bound to their neighborhood because it was impossible (until recently) to sell a house or apartment, but community leaders could have exchanged their houses and left the neighborhood through the *permuta* system, but they did not.

Most specifically, activists felt that they owe young people in the neighborhood access to recreation, play, and green spaces. Each neighborhood had a large population of youth and children that motivated residents to keep organizing and battling. As Brandy Cruthird, the owner of the Body by Brandy gym in Dudley, states: "I'm from here, I've been here a long time, and I think our kids need to see us investing in our own neighborhoods." Seeing children thriving because of improvements in parks and play spaces and opportunities for healthier living makes activists particularly satisfied with their work. In Cayo Hueso, Cristián, the organizer of martial arts classes, shares his own personal satisfaction:

I am involved because this project attracts many kids from the street. We take kids away from bad things and from being in the streets. It is a way to educate kids who have conduct problems at school. We are working with them, educating their attitudes. Their quality of life starts to improve. Their health is better.

Among the three neighborhoods, Cayo Hueso was the place that most emphasized children's well-being and health, which was reflected in projects such as the Quiero a Mi Barrio gym, the Casa del Niño y de la Niña, the martial arts classes with Cristián, the Green Map project, and the colorful Callejón de Hamel.

In some cases, parents often worked two or three jobs and children spent a lot of time outside on their own. Giving these children an opportunity to participate in sports or play in small neighborhood parks was seen as an important support structure for the whole community. Before the recent construction of neighborhood parks and playgrounds, parents who did not have time or did not own a car could not take their children to playgrounds in better-equipped neighborhoods. In Dudley, many families could not access the playgrounds in downtown Boston or suburbs like Brookline. In Barcelona, immigrant families did not let their children go to the Ciutadella Park on their own. In Havana, the intensity of car traffic scared parents away from public parks. Thanks to new green spaces, children can now play safely outside the house without a need for constant supervision.

Personal Growth and Emotional Fulfillment through Community Action

Although local activists were inspired by a strong sense of duty to participate in environmental revitalization projects, many believed that their participation in neighborhood struggles also helped them grow personally. Their work in the neighborhood fulfilled them emotionally and reinforced their attachment to their place. They were committed to caring for the neighborhood and beautifying it for themselves, other residents, and visitors, thereby helping the neighborhood grow stronger. In Boston, José Barros from DSNI explains:

We always wanted the best for the community. So if we are building a playground, we want something that lasts. We always ask for something that has value so we can feel proud of it. I envision Dudley as a place that people coming to Boston—people say, "You've got to go there."

New parks, community centers, playgrounds, and healthier housing brought a sense of modernity and progress, and activists were proud of this. They noted that local newspapers such as the *Boston Globe* and *La Vanguardia* in Barcelona and Web sites about Cayo Hueso have published articles about community projects that they have promoted or participated in.

The underlying values such as altruism and solidarity were the foundation of the work of community leaders and participants who desired to serve the community and actively transform the geographic space in which they lived or worked. Activists also expressed satisfaction at being appreciated for their work, seeing residents benefiting from it, and witnessing the community grow stronger over time. Their accounts reveal a commitment to sharing with and helping others, as noted by Mohammed, a volunteer in the Barcelona community garden, who perceived his work as a social and humanitarian service for poorer residents. Cristián, the martial arts teacher in Cayo Hueso, expresses a similar motivation:

Sometimes, some of them are a bit difficult, but if the quality of the work is good. Some kids are shy, others are more daring. They give you affection. I love the kids. It feels like you are another kid when you are with them. It is something radiant to see how they learn.

People felt the strongest about projects that community leaders and neighbors built from the ground up. Leaders such as Cristián, Rosa and Javier for the Casa del Niño y de la Niña, Jaime from the Quiero a Mi Barrio gym, and Salvador and Elias from the Callejón de Hamel in Havana believed that they realized a personal dream through their project. They used words such as *liberation, expansion,* and *energy* to describe their feelings about their work.

As activists observed the community flourishing through environmental improvements, they were further motivated to develop new endeavors and enroll others in them. This involvement built a stronger interconnection between participants over time. People got to know each other and discovered common values and interests. They also used events such as park maintenance, garden harvest, and sports competitions as opportunities to discuss new projects and social needs. The new spaces on the Forat de la Vergonya in Barcelona began as innovative environmental initiatives with playground construction, community gardens, and bike-repair workshops. Eventually, they became spaces that squatters, students, filmmakers, and architects used to show social documentaries, offer legal advice to immigrants, and organize political meetings.

Finally, the scale and space of the neighborhood allowed many participants to translate their commitment to environmental sustainability and urban nature into concrete initiatives. They were elated to realize their environmental ambitions at the local neighborhood level. For instance, in Cayo Hueso, community organizer Javier and many of the children that he worked with in environmental brigades emphasized the importance of preserving the beauty of the rare natural elements in the neighborhood and encouraged others to do the same. In all neighborhoods, people expressed how much they valued the environment in the city. Taking care of trees and plants, improving the overall greenery of the neighborhood, and maintaining community spaces made participants feel a sense of communion with nature. As Alexandria King from The Food Project explains:

I love the wilderness, I love being outdoors, I love to release through nature. And so, actually, I'd rather release into nature what I feel and have it answer back. It's a reciprocal relationship, being in the wilderness. You know that what you respect is able to give you that same respect back.

In sum, active residents, community leaders, and members of local organizations or nongovernmental organizations expressed strong positive feelings and emotions toward the fixed assets, patrimony, relationships, and history of the Dudley, Cayo Hueso, and Casc Antic neighborhoods of Boston, Havana, and Barcelona. Their accounts reflect the everyday uses and meanings of place. Activists built close ties with another another and shared common individual and collective experiences. Confluences in ways of feeling, thinking, and living in the neighborhood created common ground for neighborhood engagement. On the other hand, environmental trauma and experiences of loss and grief motivated people to take part in environmental projects. This dedication became strengthened by a strong sense of responsibility toward their place and by a desire to realize personal goals, such as community volunteering, solidarity, altruism, and enhancing nature in the city.

Healing the Community and Creating Refuges and Safe Havens

Environmental Recovery, Nurturing, and Healing

For many years, community struggles to achieve environmental recovery and improve health conditions have been a direct response to destruction, losses, and fear of erasure. People have mobilized, anchoring themselves in place attachment, built positive memories and hopes, and helped the environment recover and heal.

Green and public places humanize urban spaces that have been abandoned and fragmented, and they contribute to healing. Urban farms, community gardens, and community centers help to remediate the anxiety and trauma that residents have felt after being exposed to environmental degradation, experiencing fragile family relationships and community instability, and worrying about the uncertainty of their future. In Boston's Dudley neighborhood, The Food Project's greenhouse provides support to East African immigrants who lived through war, conflict, and loss. Staff members from The Food Project help new residents grow greens and meet with clinicians to discuss their trauma, discomfort, and fear of the unknown. Contact with nature has been linked to mental health problem prevention and treatment (Maller, Townsend, Pryor, Brown, and St. Leger 2006; Marcus and Barnes 1999). Yet places like gardens heal both individual wounds (Gerlach-Spriggs, Kaufman, and Warner 2004) and also collective ones. In Dudley, the greenhouse had a therapeutic benefit for marginalized groups. A few hundred yards away, the Boston Schoolyard initiative rebuilt outdoor classrooms and playgrounds to heal residents, soothe relationships within the neighborhood, and strengthen connections between schools and families. The new green spaces are soothing and therapeutic. As one staff member notes:

We've bridged gaps. Some neighbors had resistant attitudes, did not want new ideas in. Some have now crossed over and volunteer to water kids' garden. So we have tried to heal neighborhoods, communities, and make kids be more accepting and moving beyond wounds.

Such projects were important in the context of Boston's distress during the forced busing period. In Barcelona's Casc Antic, green spaces such as the Allada Vermell and the community garden helped newcomers from Latin America, Africa, and Asia appropriate the neighborhood as their own as much as they healed places in which older residents experienced trauma. Today, these landmarks are cherished as places where people can relax as they work on the land, contemplate trees and plants, take a stroll, and appreciate the surrounding beauty.

Environmental recovery and community healing also helped young people recover from loss and other distressing experiences. In environmental projects, participants are given new responsibilities and receive positive reinforcement. They develop an optimistic outlook. Alexandria King from The Food Project in Dudley highlights the importance of close mentoring and guidance combined with environmental work:

The team leadership curriculum is really essential to being able to process trauma. And a good deal of youth of color in Boston are suffering from trauma. . . . The key with being able to overcome your obstacles is having proper mentorship that will enable you to make the next step.

By developing mentoring activities through farms, community gardens, clean-up brigades, and sports programs, organizers and community leaders strengthened young people's mental health. In Cayo Hueso, martial arts teacher Cristián did much of his work with children who were from one-parent families and who spent a lot of time alone on the streets: "We take kids away from bad things and from being in the streets. It is a way to educate kids who have conduct problems. We are working with them, educating their attitudes." Other projects in the Cayo Hueso neighborhood, such as the Green Map, included activities in which children participated in theater or clown plays and received what their founder calls "spiritual attention" and psychological support. Community organizers saw their mentoring efforts as a way to help children process racial violence, family crisis, and school failure.

In many instances, nurturing, healing, and repairing one's place are tied to the protection of endangered traditional activities that residents engage in. People are moved by feelings of nostalgia toward disappearing practices that they attempt to recreate. In Barcelona, neighbors and environmental NGOs successfully negotiated with the city of Barcelona to create a community recycling center that was managed by a *trapero* (junkman), which used to be a traditional profession practiced in the Casc Antic. Many activists promoted urban agriculture as a way to strengthen traditional neighborhood practices, as is shown in the testimonies of organizations such as The Food Project, Alternatives for Community and Environment (ACE), the Boston Natural Areas Network, and community garden leaders in Barcelona and Havana. In Barcelona and Boston, many residents emigrated from rural regions of the United States and Spain and used to grow food for subsistence. For many, growing vegetables and fruit was a family tradition, and people expressed a desire to grow culturally valued food, such as varieties of peppers or kale that cannot be found otherwise. Today, people physically and symbolically value their fruit and vegetable harvests as a memory of their childhood. Pansy, the garden coordinator of the Leyland Street Community Garden in Dudley, discusses this deep connection to farming:

There are quite a few of people who are not from Massachusetts. They are from different parts of the country, so they decided they wanted some of their home vegetables. So they would get a spot and plant whatever they were brought up on. It's like me, I planted greens and collard greens, that's 'cause I was raised on a farm. . . . [I started working in the community garden] because I'm a farm girl.

Community gardens are one of the few places where people can have access to their culture in the city. They are inscribed in the neighborhood landscape and used as an environmental but also social and cultural resource. Community organizations support urban farming to help residents preserve their roots and remain tied to their origins. Such practices and tools heal the wounds of residents and their families. Urban farming also tells the story of a neighborhood that has been rebuilt from the ashes and is once again flourishing. It offers residents a sense of hope and tells them that the past is still present in their neighborhood. It is part of the social and cultural infrastructure of the place.

Residents who have been traumatized by loss sometimes cling to the newly created green spaces and do everything they can to defend them. Green spaces are here a matter of socio-spatial struggles among residents themselves. In Dudley, conflicts arise today between gardeners who leased a plot for years and those on the waiting list. In Cayo Hueso, the leaders of permaculture projects in housing complexes fear that passers-by or careless neighbors might harm their plots, and they organize informal neighborhood watches to ensure that others respect the new spaces. They do not want unwelcomed people to touch their plants. In the Casc Antic, the fight to preserve the green changes in the Forat was tied to protecting people from further loss. As one of the core organizers, Paco, notes:

The neighborhood was already lost, and we did not want it to be lost another time. And we thought that, with a green space and with having life right there inside, we thought that it would support this—but not if we had a hard plaza because hard plazas do not support people.

In Boston, Charlotte Kahn, a former member of Boston Urban Gardeners, explains how urban gardens anchor people in their neighborhood and give them feelings of rootedness and place: "The gardens grew out of that tumult [of desegregation]. And they really did bring forward lots of people who weren't part of, in a way, the visible landscape." In short, residents and their supporters responded to years of destruction, loss, and grief by engaging in environmental activities, which recreated a sense of placeness and reconnected them to their deeply rooted identity.

Refuges and Safe Havens

In Boston, Barcelona, and Havana, community organizations and local leaders developed environmental spaces that were safe havens and refuges. Activists helped build spaces that gave residents a sense of security and protection. They offered them warmth, comfort, and trust and responded to the emotional and psychological needs of people from vulnerable backgrounds.

Safe havens include some components of free spaces—small-scale community settings that are removed from the direct control of dominant groups and in which powerless groups are able to overturn hegemonic beliefs (Polletta 1999). They are physical and abstract places for people to be together as a group and have a warm space and refuge to socialize. They help residents remake a place for themselves in the city that is free from dominant practices and discourses about inner-city neighborhoods. These safe havens strengthen residents' capacity to deal with negative stigmas, discourses, and attitudes from outside the neighborhood. Residents sometimes feel looked down on by city officials, police agents, administrations, and outsiders in general. In Dudley, Cayo Hueso, and Casc Antic, residents had and still have daily racist experiences when they walked through, spent time outdoors in, or worked in the neighborhood. Safe havens helped them reinhabit spaces and taught them how to live well again in their neighborhood environment. Extending what Stephen Haymes argues in the context of black urban resistance (Haymes 1995), safe havens are pedagogical spaces that enable marginalized residents—including blacks in Havana and Dudley—to reclaim memories, interpret dominant definitions of urban space and its influence on how people organize identity around territory, and rebuild themselves in the neighborhood.

Safe havens include a variety of original forms and spaces. In Havana's new green street Callejón de Hamel, today Afro-Cubans practice their art and dance without feeling that they are judged or looked down on by the white society around them as it attempts to absorb them within an apparently color-blind society. The Callejón project is "a treasure that people can touch," a "resource," and an "intangible heritage," as Elias, the main promoter of the Callejón, states. Salvador himself described it as the "salvation of the cultural identity of Afro-Cubans" and as "an artistic temple" for them. The artwork, fountains, paintings, and sculptures in the Callejón are meant to help Afro-Cubans reappropriate the neighborhood spaces and recreate permanent assets. Afro-Cubans express that they can be themselves in the Callejón and that it provides them with a feeling of freedom and emancipation. It is lively and open, sheltered by the greenery and the artwork, and available for life to develop and people to thrive.

In Dudley, the Haley House Bakery Café belongs to residents. Early on, it promoted the neighborhood as a vibrant community rather than a scary place and made people feel at home. As people walk through the door of the Haley House, they often thank its founder, Kathe McKenna, for creating the place. In the café, they develop new relationships, discuss new projects, and forget stigmas, racism, and negative images associated with communities like Dudley. They also savor healthy pastries and lunches. As Bing Broderick, its manager, explains, providing such a refuge to residents is connected to environmental justice:

[The Haley House Bakery Café] was about giving people a place, a sanctuary— you know, and giving people a place to go. I think that this is related to environmental justice in a very weird way. Why shouldn't everybody have a place to go where there would be a sense of possibility and community? . . . There aren't plenty of places that are nourishing. Nourishment like on a lot of different levels, I think, is what I connect with the environmental just.

Haley House Bakery Café is isolated from the city's turmoil and is a refuge for historically marginalized groups, especially African American residents. As people read about the history of the café, attend the Art Is Life Itself discussions, or participate in cooking lessons, they reunify with their past, find a sense of release from it, rebuild themselves, and become visible again in the neighborhood and city landscape.

In Barcelona's Casc Antic, residents who rebuilt the Forat in the early 2000s converted it into a safe haven and a social space of encounter and sharing between people. Laia, a squatter and youth organizer, explains:

A social space was created in a natural manner—a space of encounter. One Christmas, people planted a tree, and around the tree, people would bring down chairs. People would sit down to talk. People from the neighborhood started to become aware that a "place to be" was lacking—a plaza, a park, a space for relations.

Such places became peaceful and secluded retreats from the city as well as places that needed to be cared for and protected. In the Casc Antic, community centers such as the Pou de la Figuera, the Palau Alos, and the Convent de Sant Agusti are peaceful spaces where residents can spend time with one another and find a refuge from the noisy city. Today, young people and families feel protected from outside eyes and police control while they play ball. Recent immigrants from Latin America, North Africa, Pakistan, and China to Barcelona's Casc Antic have been affected by racist discrimination and municipal space reclamation, but organizations such as Iniciatives I Accions Socials y Culturals (INCITA) helped youth from these countries to feel comfortable in neighborhood public spaces.

In workshops run by Silvana and her colleagues, participants reflected on how they perceived themselves and others. Short movies on their uses of space in the Casc Antic promoted their positive relationships to the neighborhood.

Safe havens include activities and programs that provided emotional support to residents and became second homes for them. Over the years, young people have participated in sports training programs, leagues, new outdoor playgrounds, community gyms, educational programs in farms, and community centers. In Dudley, multipurpose facilities have been offering physical activities and recreation programs for adolescents and young adults in a caring and protective environment. As Mike Kozu from the community organization Project Right in Dudley explains:

Safe havens are kind of like multipurpose facilities—youth centers, if you are familiar, like, with Boys & Girls Clubs. Large buildings where you can do lots of different activities for youth and community. . . . Because in the past, all they saw was the drug dealer on the street, or with the lack of jobs they felt the only way they could make money was whatever hustle, whether they steal or they do whatever street activity. We are just trying to give them positive opportunities or resources where they can make better choices.

Near Project Right, the Body by Brandy Fitness Studio recruited children who struggled with obesity and partnered with the Haley House Bakery Café to enroll participants in classes that taught healthy cooking. The gym offered a holistic vision of health, a family atmosphere, and a safe place to work out and be mentally and physically fit. Parents and instructors discussed children's progress in working out with a group. Family participation had a positive effect on the community as a whole, and adolescents developed a positive body image, strengthened their self-esteem, addressed obesity and other issues, and grew physically and psychologically.

Environmental projects provided nurturing places and refuges but also addressed violence and abuse. Early on in the 1990s, community workers in Boston and Barcelona connected their environmental revitalization projects to what they call "fights against environmental violence" and physical abuse against people from marginalized backgrounds. They saw illegal trash dumping and other kinds of neighborhood degradation as types of "common violence" against residents. Tubal Padilla, an early environmental organizer in Dudley, explains:

I articulated a relation between violence and the conditions of people, their social conditions with the environment in which they were living, and the environment as something more than simply the physical thing but also the environment of violence, and the fact that the trash was nothing else than a manifestation of this common violence.

In Dudley, destructions and dumping were systematic in the 1980s, without public authorities acting against them. Families experienced this abuse as violence against them, which affected their self-esteem. They felt that they were being treated without respect and that their exposure to toxics was being ignored by the city, especially because they were residents of color. Their mobilization around cleaning up empty land plots and developing green and recreational space brought back safety, control, and quiet to the neighborhood.

In Cuba, nonprofit organizations interpreted environmental violence differently. For them, the lack of quiet and peaceful space for children, especially girls, to play and study was a form of environmental violence. Arsenio Garcia, UNICEF's former representative in Cuba, describes Cayo Hueso before the new community projects:

[Before the projects,] there was no quiet, and there was no silence for homework. Kids with difficulties did not have the environmental conditions to fulfill their responsibilities and do their homework. There was violence against their rights, against the rights of the child, even if unintentionally.

This perception of violence against children's rights was particularly relevant in Cuba because the Castro regime concentrated much attention on children's development. The rights of the child are inscribed in articles 85, 94, 97, and 98 of the Family Code and article 40 of the Cuban constitution. Through initiatives such as the Casa del Niño y de la Niña, children could spend time in a peaceful space to play and study under the mentoring of volunteer program managers.

Finally, activists in all three neighborhoods found that socioenvironmental projects were tools for creating safer conditions despite crimes associated with guns, trafficking, and drugs. In Dudley, young people included safety and freedom from street violence as important parts of their environmental justice agenda. They reflect on the fact that constructing new parks and playgrounds is useless if those parks become the territory of drug dealers, if adolescents are shot by stray bullets as they play basketball or soccer, and if public authorities do not intervene to address this violence. Consequently, elevated crime statistics in the 1980s and 1990s led residents to advocate for enclosed community centers that offered sports equipment and activities that would have taken place in parks or open sports grounds if street violence had not scared away residents. In response to such concerns, until recently, the Boston Department of

Neighborhood Development directed much investment and funding toward enclosed centers rather than open and clear spaces.

Green space reclamation also can create a buffer against gun- or drug-related crime. In Boston's Dudley neighborhood, residents reclaimed abandoned open areas and transformed them into active sports grounds and green space to deter criminal activity. One well-know victory is embodied by the residents's occupation of the Mary Hannon Park in 1993. In that summer, young people and families organized recreational activities, sports leagues, and clean-up events to push away drug traffickers, deter drug sales, and act as a barrier against violence. This occupation and the activities taking place in the park sent a signal to drug dealers, created new uses for the space, and encouraged the city of Boston to invest municipal funds and renovate it the next summer.

In Cayo Hueso and Casc Antic, where violent crime was uncommon, community leaders concentrated their efforts on discouraging drug use and trafficking through neighborhood clean-up workshops and sports activities. Sports activities occupy the minds and bodies of adolescents after school and encourage them to focus on activities that positively affect their development. In places such as Quiero a Mi Barrio in Havana, Javier supervised and coached adolescents after school in a variety of physical training activities. Project supervisors also monitored youth educational achievements and discussed the danger of drugs and crime with young people. In Barcelona, the coordinators of the A. E. Cervantes–Casc Antic (AECCA) sports association, the Fundació Adsis, and the Fundació Comtal supported basketball training programs and sports leagues to help keep children and youth off the streets and away from undesirable activities. Today young people play and interact in an environment where they are mentored and can use their energy in a positive way.

In sum, activists in Cayo Hueso, Casc Antic, and Dudley used environmental and health projects to address loss, create safe havens, and remake a place and home for residents. Through their projects, they addressed multiple dimensions of safety (figure 5.2).

Celebrating the Neighborhood and Strengthening Local Identity

Activists who helped to develop refuges and safe havens also viewed those new spaces as opportunities to celebrate and honor the neighborhood, strengthen a positive individual and collective local identity, and create a foundation for future action.

Figure 5.2

Environmental justice, safety, and place remaking in environmental revitalization projects

A Self-Sustained Urban Village

As local leaders and workers in neighborhood associations and organizations revitalized the Dudley, Cayo Hueso, and Casc Antic neighborhoods of Boston, Havana, and Barcelona, they also attempted to recreate the life and conditions of an urban village. They shared a nostalgia for the welcoming and familiar place that their neighborhood used to be. An urban village is not only embodied in close-knit groups of families living in narrow winding streets full of urban social life (Gans 1962). It is also a self-sustained entity with multiple uses of space, compactness, a vibrant life, and a range of work and leisure activities. In Havana, the Green Map project designed self-sustainable communities that "mobilize local resources and efforts." In Dudley, an urban village put into practice a vision for urban sustainability. As Greg Watson, former DSNI executive director, notes:

Roxbury was a place. It was a retreat from the city. We said, "Let's bring back some of the native, . . . , the cultural, aspects of a multicultural neighborhood that could be used to create the sense of place that would have low environmental impact but could also contribute to economic development. We would throw in the urban agriculture."

Recreating a village unites residents around a dynamic, protective, and vibrant place. An urban village also includes new community landmarks for people to connect to, such as Dudley Commons in Dudley, the Pou de la Figuera and Allada Vermell spaces in Barcelona, and the Callejón de Hamel and Parque Trillo and Beisbolito green spaces in Havana. Residents are able to walk to a variety of stores and services in the urban villages, and develop organic and low-impact farming on site, especially in Cayo Hueso and Dudley. Community organizations such as The Food Project, ACE, DSNI, and the Boston Natural Areas Network promoted urban farming and community gardening as a way to create an environmentally sustainable village while fostering a cohesive community in the middle of Boston. As a reinforcement, throughout the development of new environmental projects in Dudley, leaders such as Powanie Burgess intervened in neighborhood meetings and promoted the identity of the place as a farming community.

Day after day, an urban village is built by organizing traditional and communal work activities such as clean-up events and community garden maintenance. Volunteer work benefits the whole community. Garden and farm harvests were shared with neighbors during social events organized in Dudley and the Casc Antic. Much of Barcelona's Forat reconstruction in 2000 and 2001 started with festive events around organized paella cookouts where musicians and clowns entertained participants who worked in the gardens. Two hundred people spent all day together preparing dishes, planting new crops, and building new playground equipment. People enjoyed the gathering and did not feel that they were working.

In many instances, community work on environmental projects actually became a pretext for residents to gather outside and reconstruct relationships with their neighbors. In large and impersonal cities, environmental projects have helped residents reinitiate and rebuild the close human rapport that exists in small villages. People expressed a desire to be together. As Monika, a squatter who was active in the Forat activities, explains:

It was the desire to be in a place where there was life, community, and neighborhood life because it was one of the things that Barcelona had a lot but has much less now. But it has neighborhoods where people know each other and do things in the streets, and the kids play in the street.

Behind such expressions of community life lies a desire to reappropriate physical and public spaces in the neighborhood. Family pastimes extended into streets and parks, as was illustrated by signs that were hung from the balconies of the Casc Antic: "Els carrers com a casa nostra" (The streets are like our house). In Cayo Hueso, community-building events often took the form of cultural celebrations called *peñas*, which frequently occurred during the remodeling of *solares* buildings. These *peñas* were organized by a now-deceased blind community artist, Eladio, around singing, dancing, and rumba festivities. Minivillages were created in the *solares*, where residents worked together to repair and maintain their buildings while participating in cultural events and enjoying each other's company.

As projects brought people together, families joined the activities and became more integrated with each other and in the neighborhood. In the Casc Antic, the sports organization AECCA involved parents in the project and made them a part of its development rather than customers or clients. As a staff member points out:

There is a tendency that we want to try to eliminate—that families stand up for the project only when they register their kids—because we are also interested in getting them to feel integrated, not as a social service that we would deliver but rather that they form part of it. AECCA is not me or Montse. It is all of us.

This form of thinking was also illustrated in the vision behind the new Kroc Center's community gym in Dudley. The facility includes a Village Center that welcome people to the gym, offer them meeting and café spaces, and invite families to relax and chat with each other. The Village Center is both a place of passage and a gathering space. The Kroc Center encouraged residents to build closer informal relations with each other as they exercised or played with their children in the new pool.

Community Celebration, Sharing, and Learning

Place remaking cannot be separated from celebrating the diversity of residents' ethnic and social origins and making them more visible in the urban village. Multicultural events and festivals are common experiences that activists organize to rally residents around neighborhood projects, showcase the community, and strengthen ties between them. Planned around community gardens, park clean-up, and fundraising rallies, these festivals and celebrations are occasions for residents to build connections in a cheerful and friendly atmosphere. Environmental revitalization becomes a tool for community celebration.

Every August in Dudley, DSNI organizes a multicultural festival in Mary Hannon Park to highlight the community's diverse population and strengthen residents' cultural identity. The event includes arts and crafts activities, skating performances, a youth fashion show, sports competitions, and food representing many cultural and ethnic groups. Every summer, the Boston Natural Areas Network sponsors a series of neighborhood concerts in Dudley gardens. Such events bring together gardeners, celebrate accomplishments, and strengthen the tighter community in a festive atmosphere. In addition, every week during the year, the Haley House Bakery Café organizes events around art, history, and culture (such as Art Is Life Itself) that attract residents to the bakery and café. With the help of local artists, historians, and community leaders such as Mel King, the art series celebrates the neighborhood and its diversity, addresses internalized oppression from racist discourses and attitudes about the neighborhood, and fosters people's interest in eating and cooking healthy food.

In Barcelona, the construction of new community gardens was accompanied by the celebration of residents' traditions. Volunteers prepared couscous and Moroccan tea for celebratory parties that were held in the gardens. Today, the new leaders of the community garden in the Pou de la Figuera bring in children's cultural roots and traditions into the project to make them feel more at home in their new neighborhood and teach other residents about their neighbors' traditions. Maria, an active garden volunteer, includes plants from Africa and gardening practices from the Caribbean. Next door to the garden, the community organization Mescladis offers healthy and organic dishes from several parts of Africa and share residents' cultural roots with clients.

In Cuba, culture, art, and music permeate the daily life of every citizen. Gathering around cultural and artistic events is a natural way for people to come together and escape daily hardships. In addition to being a space for youth recreation, the Casa del Niño y de la Niña promoted local artistic talent. Many of the activities planned at the Casa showcased children's artistic talents through painting and salsa. On the other hand, a GDIC staff member explains the importance of cultural traditions and events for attracting residents to new projects in Havana's Cayo Hueso neighborhood:

It was the easiest to come in and do work in the community—both to use culture and identity. You are not imposing anything to the people. . . . You create a canal of communication and trust. You create a gathering of people—for instance, in sports activities. From there, you can talk about health, environment, and education.

Today, in the Callejón de Hamel, the Sunday rumba events welcome visitors from the whole city and bring together residents in a carnival-type atmosphere. For Salvador, its founder, the ultimate vision for the project was to reintroduce Afro-Cuban art and dances into the neighborhood and present Afro-Cuban traditions to the outside public. Some of the environmental improvements that he sponsored—such as infrastructure and sanitation repair and the greening of public spaces—were indirect tools for convincing authorities to leave him alone as he developed his project's cultural and identity aspects.

Several community initiatives highlighted the multicultural traditions and artistic talents of neighborhood groups, but they also encouraged learning from one another. Community gardens in Dudley and Casc Antic have created intergenerational exchanges among children, young people, and elderly residents who had farmed for years. Young residents remain connected to their origins, as Alexandria from The Food Project explains:

> For people who are coming over here and their descendants, there is this disconnect because you have the "What is it?" Becoming assimilated into American culture, so you kind of lose your roots. For a lot of the elders in the community, things like farming are second nature and key. But now young people can kind of see the connection to that and respect it and get involved.

Through farms, local youth develop stronger ties to the older community members and to their traditional practices.

As residents learn from one another and share with their neighbors, they also learn to build consensus between points of views and eventually achieve change together. In the Casc Antic, staff members from the GENAB environmental NGO noted that environmental education on waste management and recycling addressed tensions between residents, improved interpersonal relations, and enhanced coexistence. Residents of high-density buildings used to fight with each other over the mishandling of waste and lack of care for the buildings. As residents became more environmentally and socially aware, they also developed a greater sense of responsibility toward their neighborhood. In Cuba, Rosa from the Casa del Niño y de la Niña helped residents use environmental brigades and garden work to transmit values such as respect for and responsibility to others around them.

Addressing tense relationships within the neighborhood was important because clashes occurred between younger and older residents and between different cultural and ethnic groups. Environmental justice projects were and are still used to enhance coexistence and understanding between people from different origins. Sports leagues in Dudley, Cayo Hueso, and Casc Antic brought youth groups closer together. Principles of tolerance and openness are reflected in mission statements, such as

AECCA's: "Through the practice of basketball, the needs of the different present cultures have been united, while creating a real atmosphere of coexistence and especially transmitting diversity as a value." Sports help players to build positive attitudes toward one another, erase preconceived judgments, and leave disagreements off the court. When children from different origins bring their parents to watch them play, their open attitude helps decrease the mistrust that adults often have toward one another, especially in multiethnic neighborhoods.

Not only did cooking classes at Mescladis in Barcelona and at the Haley House Bakery Café in Dudley teach students how to eat and make healthy and affordable dishes but they also provide lessons about cultural openness, mutual help, cultural and racial stereotypes, and cooperation. Children combine ingredients from different and unknown cultures to make tasty recipes. At the Haley House, Deedee, chef and teacher, explained how ingredients from different cultures symbolically represent neighborhood groups and how their presence enriches the fabric of the place. She uses healthy food as a way to cross racial, cultural, and economic divides and to open children's minds about others. Participants revise their preconceived negative opinions, and the community become more resilient in dealing with conflicts, differences, and threats of division. In Barcelona's new community garden, children told stories about the ways that certain plants and vegetables were used in their home country or region. Through storytelling, children recovered their past while experiencing a sense of reunion and liberation.

Despite efforts to bring residents together, tensions remained at times. In the Barcelona's Casc Antic, some groups, such as the Latino and Moroccan participants, had difficulties mixing with one another in the sports leagues. In Boston's Dudley, conflicts around the management of some community gardens led some groups to secede from the original group and create their own garden. Women gardeners from the Leyland Street Community Garden started a new garden a few yards away from the original site. In Havana's Cayo Hueso, Afro-Cuban residents felt that the government might close the Callejón de Hamel project or the Casa del Niño y de la Niña in response to the lobbying by white conservative Cayo Hueso residents to the neighborhood councils and the Comité de Defensa de la Revolución, arguing that such projects are counterrevolutionary.

Those clashes reflect tensions that arise between encouraging cohesion and homogeneity and also celebrating its diversity. They also illustrate the tensions between cooperation and competition between groups. A collective neighborhood identity cannot erase differences and individuality, which is why local leaders and organizations work to rebuild a collective identity that is based on shared lived or represented history about loss and recovery and environmental revitalization accomplishments. These are the delicate balances that community organizations have to work with to ensure continued engagement and participation in local projects.

Memory and Collective Local Identity

As activists work to create an urban village and celebrate community groups, they attempt to construct a common memory and identity. Maintenance of community gardens, sports competitions at a new facility, and art shows in healthy cafés and along green streets are often accompanied by historic lessons about the neighborhood. Community organizers and leaders hope that, as residents share a territory and experiences and build a common place memory, they increase their sense of belonging, feel more identified with their immediate community, and eventually participate in neighborhood revitalization projects. In Boston's Dudley, Tubal Padilla, an early environmental activist, explains what he sees as a logical path:

If people do not have a strong sense of belonging, if people are not identified with their immediate communities—well, they will not participate. . . . To identify this place as the Dudley Street neighborhood is something that many people were not doing before. But it was a transformation of identification of the ethnic groups who shared a territory. So on the one side, there were the common enemies and, on the other hand, a common project that took much time to develop and to recruit the vast majority of people.

Unifying people around the idea of a community that exists beyond each ethnic group was a challenge for Dudley, but it eventually brought a broader support base to activists.

Throughout the years, community organizations—such as Dudley's DSNI, ACE, Project Right, and Discover Roxbury; Casc Antic's Hortet del Forat and Veïns in Defensa de la Barcelona Vella; and Cayo Hueso's TTIB workshops—told many stories to residents about the neighborhood's former degradation, abandonment, and revival. They wanted to remind them of the importance of working together and sustaining recent achievements. This kind of memory construction is an important rallying tactic for building a common understanding of abandonment and reconstruction in the neighborhood. Trish Settles, a former environmental organizer for DSNI, emphasizes the importance of memory:

Being able to link all those different pieces of community, draw from those connections not only for themselves but also for the funders (and say, "Hey, look, we need the resources") and for the residents in the neighborhood. The residents have a common understanding of the kind of the issues that are out there: "Well, yes, we do need housing. Yes, we do need to protect our open spaces. Yes, we do have a soil contamination problem. But we can address these if we all work together."

Specific events can help people learn about past battles and accomplishments. The 2.5-mile Walk for Dudley takes place every October and celebrates the community-based organizations that worked to shape the Dudley neighborhood into a vibrant urban village. Throughout the walk, organizers lead participants through the Kroc Center, The Food Project farm, and renovated parks. As long-term political leader Mel King explains: "That march is a way of involving people, strengthening and carving out ways of making some demands on each other."

As participants exchange and share common memories, they build a positive collective identity. They realize that past accomplishments need to be protected in the long term. In Barcelona's Casc Antic, gardeners regularly shared stories about neighborhood fights in 2000 to 2007 and about the courage that residents demonstrated against the municipality's development plans. They hung photographs of fights between residents and the police around the fence of the garden so that neighbors kept a visual memory of the recent past. Today, garden volunteers and community organizers still organize evening meetings in the community center to explain the local environment and development changes to participants. Narratives about the past build a symbolic connection to community struggles and changes for residents, especially for newcomers who might not know the history of the neighborhood. In a similar manner, in Havana, the TTIB workshop staff members spent much of their time teaching the history of Cayo Hueso. Joel, the former TTIB coordinator, feels that residents need to learn more about their place to then be motivated to take care of it:

Cayo Hueso has such as long cultural and patriotic history. We are constantly remembering it here in the TTIB: it is important that the residents know their history and that they love their neighborhood and that they feel the neighborhood. It is important to develop the feeling of belonging and the sense of taking better care of the neighborhood and social relations.

Within each neighborhood, some community leaders fostered and still foster today strong feelings of belonging and encouraged residents to play a role in neighborhood revitalization. They often are older residents with a vivid knowledge of neighborhood history and strong oral skills that attracted participants. In Havana's Cayo Hueso, Rosa from the Casa del Niño y de la Niña led many history workshops. Her enthusiastic voice and animated body language easily drew in participants. As children and adolescents participated in her workshops, they developed a collective memory of the place and built a durable place identity. As Rosa emphasizes:

I have a strong sense of belonging. I give workshops on the history of the neighborhood so that they love their neighborhood because it is important to take care of the neighborhood. I explain to them the context, the history. So that makes them feel more committed, and they behave better, and they take better care of the environment.

Rosa created a competition called Mi Barrio (My Neighborhood) so that children could talk about their neighborhood (what they like about it and would like to protect in it) and draw parts of it. For her brigades Por un Barrio Más Limpio (For a Cleaner Neighborhood), she gave participants identity cards designed by Pablo so that they felt proud of their work and had a tangible memento of their participation.

Finally, art has been a valuable tool for representing past struggles and accomplishments on neighborhood streets and buildings, which in turn strengthens residents' attachment to the neighborhood. Murals are a traditional expression of street art in communities of color, and they help residents reclaim their roots. In Boston's Dudley in 2009, a group of youths spent several weeks on East Cottage Street working in a summer jobs program to research and paint a mural that celebrated the Dudley Street reclamation process as a story of arson, loss, and resurrection. As Travis Watson from DSNI describes:

One example is a mural that was recently done by a youth group this past summer that was housed here at DSNI. It's on East Cottage Street, and it kind of shows the history, briefly, of the community, of DSNI, of the Roxbury, Dorchester, Upham's Corner community. So there is always a reminder everywhere you go of "Oh, yeah. It's still alive"—that sense of community.

Murals are part of local history. They capture the richness of the place and its role in the city and pay a visual tribute to the lives of residents. As murals showcase local artistic talents and beautify public spaces, they also represent the energy of the community and increase proximity between people. Discussions of neighborhood events and features to be drawn on the murals encourage collective reflection, unity beyond difference, and peace, and eventually they build a common neighborhood identity without negating differences and individuality.

In sum, local residents and their allies used environmental revitalization projects to create a closely knit village, celebrate the diversity and history of the community, enhance mutual learning and social coexistence, and recreate a strong and affirmative local collective identity beyond a past of decay and abandonment.

Discussion: Environmental Recovery as a Tool for Remaking Place

In this chapter, I have looked at how place and neighborhood-based activism intersected to help transform environments, how place was influenced by the local political economy and urban developments, and how activists responded to changes and disruptions in their neighborhood. Place plays a dual role because it both motivates actions and provides goals about place remaking. It also has a dual meaning that is related to both negative and positive emotions—both grief and hope. Activists and their supporters have memories of suffering and loss that are related to place but also positive memories, and both shape their relationship to their neighborhood and their engagement in it.

The stories of activists in the Cayo Hueso, Casc Antic, and Dudley neighborhoods show that place attachment and sense of community motivate residents to organize long-term environmental revitalization in distressed neighborhoods. Residents and their supporters have a strong connection to the neighborhood's physical and architectural features and the relationships that they have built there (Scannell and Gifford 2010). This connection also rests on a neighborhood's history and traditions, on common experiences of fragmentation and disruption (Massey 1994; Fullilove 2004; Campbell and Fainstein 2003; Sandercock 2003), and on activism.

Residents' experiences of the place also allowed them to realize the pervasive consequences of neighborhood abandonment, decay, and degradation on environmental quality and local identity. Marginalized neighborhoods are imbued with negative experiences and environmental trauma. Disruptions forced people to acknowledge and address the question of place. This double experience of place attachment and abandonment—strengthened by a sense of responsibility, a desire for personal growth, and a commitment to the urban environment—prompted activists to become engaged in their locale through socioenvironmental projects. At the local level, broad differences in levels of development and democracy do not seem to affect how individual and groups interpret exclusion and sense of place and community as motivators for action.

However, place is not only a motivator for local action (Chavis and Wandersman 1990; Cohrun 1994; Davidson and Cotter 1993). Neighborhood-based environmental work remakes a place for residents and helps the neighborhood flourish from within. As activists repair community spaces, build new parks and playgrounds, and develop urban farms and gardens, they address grief, fear of erasure, and suffering in a neighborhood that they once saw as a war zone. Projects such as green spaces and community gardens repair a broken community and heal both individual (Gerlach-Spriggs, Kaufman, and Warner 2004) and collective wounds. They contribute to the environmental recovery of sites and also groups. Environmental recovery has territorial and physical dimensions but also social, mental, and human components.

In urban neighborhoods, environmental justice and safety are closely connected because environmental projects create a greater sense of safety for residents. This sense of safety can help residents overcome grief and environmental trauma, recreate feelings of rootedness and nurturing, create and protect refuges, address mental insecurity and negative images about themselves and the neighborhood, reappropriate the neighborhood geography, deter environmental violence and dumping, and prevent crime in the neighborhood. Urban environmental justice encompasses aspects of safety that go beyond protection against physical, social, and financial damage to incorporate cohesiveness, wholeness, soothing, and protection. Many new environmental spaces are safe havens. They are physical and emotional spaces for marginalized residents, especially residents of color, to rehabit the neighborhood territory, become visible again, reclaim memories, and redraw positive connections to their place. As people engage in environmental projects, they rebuild themselves.

Finally, community leaders, neighborhood workers, and staff members of local organizations use environmental projects as tools for creating an urban village, which is embodied by a tight and traditional social fabric, traditional farming, recreational practices for migrants and immigrants, and opportunities for socializing outdoors. Projects also celebrate the neighborhood's diversity, address internal conflicts and divisions, and rally residents who were not previously engaged in neighborhood action. Initiatives such as housing and sanitation improvements, garden construction, and community environmental planning include activities that build a positive collective memory and identity.

Some nuances exist between neighborhoods. In Havana's Cayo Hueso, activists have not experienced what others in Boston's Dudley and Barcelona's Casc Antic have called urban war, but they still connect the loss of neighborhood assets and goods to urban-renewal projects and permanent crumbling buildings. In the Casc Antic, expropriations and building demolitions in the 1990s traumatized residents and led to a massive exodus, while loss and trauma in Dudley was connected to arson, street violence, and neighborhood degradation. Activists in Dudley also confronted crime and violence, which shaped much of their environmental revitalization work. In Cayo Hueso, Afro-Cuban leaders connected their fear of erasure to the Castro regime's attempt to assimilate them into a white socialist mentality. Race mattered in Dudley, as well, and has mattered more in the Casc Antic in recent years. Race often is the basis for cohesion inside the neighborhood and also for feelings of exclusion and threats outside.

In these three neighborhoods, activists' struggles reveal that both physical and psychological dimensions of environmental health must be considered if marginalized neighborhoods are to be holistically revitalized. In neighborhoods such as Dudley, Cayo Hueso, and Casc Antic, urban and just sustainability (Agyeman, Bullard, and Evans 2003) becomes enriched with social dimensions that are not limited to poverty alleviation and social justice. Sustainability involves reconstructing place, alleviating trauma and loss, nurturing, celebrating traditions and cultures, and enhancing coexistence. Communities become more resilient and robust as activists integrate wellness with environmental revitalization. Activists attempt to create a form of social sustainability that protects and strengthens the social and human infrastructure of the place.

The symbolism of urban farms, parks, and community centers is distinct from the physical improvements achieved by neighborhood advocates. Community struggles remake a Gemeinschaft in which residents are bound by cultural ties and values, collaborate together, and show altruism. The work of residents and their supporters reveals the tension between creating a place as a material reality and using place as a repository of emotions and memories and a vision for the future. Here, activists can be both long-term residents and newly arrived migrants who feel that they have a stake in their new neighborhood and want to make it their home. Their struggles demonstrate that contemporary global cities do not lose their accumulation of shared history, memories, and engagement with public spaces under threats of privatization (Sassen 1998).

As activists in Boston, Barcelona, and Havana became aware of the degradation, losses, and possible erasure of their neighborhood, they did not operate in isolation. Their rebuilding work (re)-created a place for residents, but also confronted broader urban economic and social processes in which the neighborhood was embedded. Without those broader political agendas, community rebuilding and place remaking cannot be sustained. The analysis of the broader political agendas framed and defended by local activists is the point of the next chapter.

6

Advancing Broader Political Agendas: Spatial Justice, Land and Border Control, and Deepening Democracy

When activists of the urban communities of Dudley, Casc Antic, and Cayo Hueso decided to fight for environmental and health improvements, they did so because they were attached to their neighborhoods, had a vision for place remaking, and wanted to overcome experiences of grief, loss, and trauma. Yet, their mobilization was also situated within a broader context of urban political and socioeconomic processes in Boston, Barcelona, and Havana that affected neighborhood stability, dynamics, and development. In this chapter, I focus on the extent to which the environmental struggles of marginalized communities advanced larger political agendas in the city and contributed to changes in political processes and outcomes.

In Boston, Barcelona, and Havana, community activists have been politically involved in other neighborhoods, other struggles, and national politics. Their support for residents in Dudley, Casc Antic, and Cayo Hueso was a natural and logical outcome of years of political activism in radical movements and territorial conflicts. For instance, activists in the Casc Antic came from organizations and groups that fought against uncontrolled tourism, real estate speculation, and expansionist urbanism throughout Barcelona. There is thus no separation between environmental goals and political goals. People feel satisfaction when they become part of local political actions and part of a movement that includes a wide variety of protestors. Many of their daily activities in the neighborhood have a broader political significance and dimension.

The stories of how activists became involved in environmental revitalization struggles reveal that many centered their lives around their community work and some could not imagine themselves living without fighting for others. The words they used to describe their engagement suggest that they saw themselves as part of a political trajectory. Emanuela, an Italian supporter of the Forat struggles in Barcelona, shares her own interpretation of community engagement: "For me, it was part of my trajectory—to involve myself. I also liked the idea of doing a garden. I was living a trajectory."

Activists in these three neighborhoods took pride in telling lengthy stories of their battles and their victories despite injuries and arrests. They believed that the neighborhood chose them rather than the opposite. In Havana's Cayo Hueso neighborhood, people considered themselves to be at the avant-garde of political changes in Cuba. As Salvador from the Callejón de Hamel says: "I feel that I have a great responsibility because I took this on in 1990, when it was still a taboo to express oneself in such a way."

In this chapter, I argue that activists cannot sustain community rebuilding and place remaking over time without working for broader political agendas and goals. Environmental justice struggles cannot be fully understood and achieve long-term success without an examination of the broader urban political economy of environmental inequality production. Table 6.1 summarizes the broader political agendas that activists developed by identifying negative influences on their neighborhoods. Environmental projects became tools to address racism and vulnerability, ensure spatial equity, and recover residents' right to the neighborhood. They also allowed activists to control land and borders and achieve a form of spontaneous collective participation.

Opposing Racism, Classism, and Vulnerability

Changing Images and Discourses about Neighborhood

Distressed urban neighborhoods often carry stigmas that are associated with substandard housing conditions, degradation, violence, and poverty (Falk 2004; Gotham 2003; Gotham and Brumley 2002; Manzo, Kleit, and Couch 2008). The neighborhoods of Dudley, Casc Antic, and Cayo Hueso are no exception, and people there developed negative images about themselves, their socioeconomic conditions, and the physical environment. In the 1990s, outsiders called these areas urban ghettos, which they saw as plagued with unemployment, decay, and violence. Public officials and media sources considered them social purgatories where no one but social outcasts lived. Prejudicial beliefs about them limited the creation of new opportunities for residents and reinforced their marginality (Wacquant 2007; Garbin and Millington 2012).

Territorial stigmatization came from a variety of sources. Municipal reports and press releases highlighted dramatic states of disrepair, social

Table 6.1

Broader political agendas in environmental revitalization projects in three neighborhoods: Dudley (Boston), Casc Antic (Barcelona), and Caya Hueso (Havana)

Negative practices and threats	Target	Political agendas	Means and strategies
Stereotypes and stigmas about the neighborhood	Media reports, reports from municipality, public opinion	Fighting racism, exclusion, and vulnerability	New practices and discourses highlighting positive neighborhood transformation and combating stigmas and stereotypes
Urban developments	Private speculators, planners, public officials	Right to the neighborhood; combating encroachment and environmental gentrification	Affordable housing advocacy; use of spatial equity lens; emphasis on publicly used space and commons
Outsiders and newcomers	Gentrifiers, tourists, young urban professionals	Controlling land and borders	Appropriate tenure or lease mechanism; choice of specific design for environmental projects; social and cultural markers of difference in projects
Coopted participation, traditional deliberative and representative democracy, government control	Public officials at the local and national levels	Fostering spontaneous collective participation and rebuilding local democratic practices	Learning spaces in new environmental commons; use of bricolage techniques

problems, and an unwelcoming atmosphere. These connotations originated in the history of each neighborhood as it declined into a crime-
ridden, valueless, abandoned neighborhood that was seen as populated
with underclass citizens. After decades of decay, white flight, and disinvestment, outsiders considered Boston's Dudley neighborhood as a place
to avoid. As Gregory C. Watson, former director of the Dudley Street
Neighborhood Initiative (DSNI) and current commissioner of the Massachusetts Department of Agricultural Resources, explains: "If you go and
get tourist maps, there is a little black hole right around that particular
neighborhood. They tell you to avoid it. I'll never forget when I got there,
and we got the Orange Line [train] maps and directions, and they would
say 'Avoid.'" A current Dudley leader adds: "The media only want to see
the badness. They don't want to show the goodness."

To a similar extent, residents in Barcelona's Casc Antic neighborhood
remembered how the city long considered the neighborhood to be a ghetto and a place that needed rehabilitation. Media reports reinforced this
impression. In 1998, several local and national newspapers highlighted
the concentration of immigrants in the neighborhood: "The foreign population has multiplied by three in ten years in the Ciutat Vella" (*Avui*, September 1, 1998); "Ciutat Vella concentrates the foreign population living
in Barcelona" (*El Pais*, September 2, 1998); and "23 percent of children
born in the Ciutat Vella are from immigrants" (*El Mundo*, September 9,
1998). Those publications chose to share a story about the concentration
of foreigners in the Old Town as the most salient message about immigration in the city. Such details indirectly indicated to readers that the
place where immigrants resided was the most important variable about
immigrants and that they resided in a ghetto. These newspaper articles
followed multiple reports on the concentration of immigrants in the city,
such as a report titled "Local Administrations Facing Migratory Facts,"
which stated:

Low economic productivity and work instability, tensions that stem from a limited ability to adapt to our customs and a certain distrust of some sectors of the native population about renting housing, create a concentration [of immigrants] in
the most deteriorated urban areas, which, in certain cases, have become ghettos.[1]

According to the report, ghettos were a reality in which many social bads
needed to be addressed.

By engaging in environmental and health projects, community activists
in all three neighborhoods addressed negative stereotypes that were in-
fused with racist connotations. As revitalization projects transformed the

neighborhood and became new landmarks, they slowly helped to change its image. In Havana's Cayo Hueso, Cristián Rendon, the martial arts teacher, explains the vision for his work: "To change the image of the neighborhood is one of the principles underlying our work. It is a question of education and training. We show that it is neighborhood of people who are only a family."

Similarly, in Boston's Dudley neighborhood, activists showed that environmental issues are relevant for black and Latino residents. Alexandria from The Food Project explains: "You are definitely breaking down stereotypes. And I think some of them are that people of color don't care about the communities that they are living in—that people of color don't care about their health." Changing stigmas about Dudley and its residents was particularly important because minority populations are at risk for obesity and cardiovascular disease and often are portrayed as not taking care of themselves. Urban gardens and farms thus played an important role, as the coordinator of the Savin and Maywood Streets Community Garden highlights: "The press puts out a lot of things about the neighborhood that are wrong. Gardens help change the image of the city." A few blocks away, in places such as The Food Project farm, suburban youth worked with inner city youth residents during the summer. As they cooperated and shared responsibilities around the farm, they discussed negative biases and revised their own preconceived opinions about each other.

Beyond contesting stigmas and negative beliefs about each place, activists attempted to confront racism toward residents. They emphasized that their work is a means to resist "internalized oppression," "institutionalized racism," "cultural racism," and "interpersonal racism." For Alice Gomes, a former environmental organizer in Dudley, the illegal dumpling that residents suffered without any intervention from public authorities was grounded in racism:

It was racism. I mean, what other community can they go in and do what they did in Roxbury back then and get away with it? It was racism, and I think it was classism also. Because again it was a working immigrant community, and people were trying to make ends meet.

Today in Dudley, the founder of the Haley House Bakery Café feels that the new café breaches racial and economic divides in Boston as clients and visitors connect and develop mutually beneficial relationships. In Barcelona's Casc Antic, social and cultural events organized around the Mescladis food project and community garden bring greater visibility to Arabic, Latin American, and Indian cultures. They facilitate the training and hiring of immigrants, who generally encounter legal barriers and racism as they seek jobs and professional development in the city. Food can be used to transmit the values of multiculturalism, respect, and diversity.

In Cayo Hueso, some of the initiatives that residents led addressed subtle manifestations of racism and marginalization that Afro-Cubans have suffered. In Cuba, the revolution of 1959 led to a desire to construct a new national identity and a new definition of *Cuban* that would supersede any other. As a result, the identity of Afro-Cubans as blacks was diluted in the push for integration (Morales Dominguez 2004). The issue of race was eliminated from public debate because it was considered to be counterrevolutionary and divisionist (De la Fuente 2000; Hernández, Vásquez, Zardoya, and Mejiles 2004). Consequently, many of the racist ideologies that people held when Fidel Castro took power were displaced to private spaces and were displayed in daily life through small acts of exclusion, such as the use of racial labels and the exclusion of Afro-Cubans from some private spaces throughout Havana (De la Fuente 2000). In Havana, being marginal and being marginalized has been closely related to racial prejudices against Afro-Cubans, which for a long time, were a taboo topic in Cuban politics and culture (Hernández et al. 2004).

In this context, some of the projects that were developed by Afro-Cuban leaders in Cayo Hueso were attempts to overcome the domination, alienation, and paternalistic attitude of a national culture that rejected alternative Afro-Cuban behaviors as secondary, subordinate, and disturbing. Afro-Cuban voices have started being heard and legitimized, and Afro-Cubans want to participate in the making of Cuban society through their practices. Elias from the Callejón de Hamel explains how the artist Salvador tried to construct a racial democracy by changing mentalities and transgressing the standards of the official white Cuban culture:

Here we promote the Callejón de Hamel in a different way from what the Cuban government does. . . . In general, many people considered the rumba as something pornographic. The fact is that religion did not fit with the image of the Che [Guevara] and with the atheist regime. . . . You just don't take away taboos here. There is a strong preconceived opinion against religion and against race. [The artist Salvador] brought about this transformation and transgression on his own private initiative. . . . The state built a stereotype around the Callejón de Hamel. In Cuba, the open street festivals were prohibited because African religion parties were felt to promote the darker side of life.

Projects such as the Callejón enhanced the value of the neighborhood for the city and the broader culture. Beyond the improved environmental conditions that the project brought to the neighborhood, it also eroded negative views about Afro-Cubans that were still held in Cuban society.

Addressing Vulnerability and Offenses to Dignity

Through community projects, activists addressed stigmas, negative images, and racist attitudes toward the neighborhood and its residents. These projects also helped them address broader dignity and vulnerability issues that are connected to classism. Residents, community organizers, and neighborhood leaders argue that families and individuals deserve to be treated respectfully and not as second-class citizens who were born to live in crumbling buildings without access to a clean environment. In Boston's Dudley, Father Waldron explains the broader rationale behind his engagement:

For me, it was a people thing, not an EJ thing. Environment was not mentioned at first. It was poverty right in your face. Later, people saw a connection with housing and healthy living. "Don't dump on us" did not have a strong environment focus. People dumped because this area had no clout. "You are using our neighborhood." People came to the realization of abuses.

Residents who live in poor neighborhoods are as valuable as those in wealthier neighborhoods, and community leaders and their supporters vowed to defend their rights. In Cayo Hueso, nongovernmental organizations from outside the neighborhood initiated their work as part of an action of solidarity toward the most vulnerable and isolated residents in Havana, as Joel from the Martin Luther King Center explains:

Our insertion in the community/neighborhoods came from our motivation to show solidarity. It was a space for activities in 1987. With the crisis in 1990 came more challenges and emergencies. . . . Our mission was to accompany prophetically the people of Cuba. This was also embodied in smaller projects. The need was the sense of a community in the city of Havana. *Accompanying* meant living the process with [the people], not giving them orders. . . . It was a seed to help solve problems of inequalities in Havana.

Organizations such as the Martin Luther King Center addressed growing social injustices in a city in which there was no space to debate these topics in the first decades after the revolution.

In some cases, the municipality and the decisions that some offices and public enterprises took in the neighborhood were a direct target of residents' fights. Some activists accused public officials of taking advantage of residents as powerless victims of abuses and of not protecting equally everyone's right to a safe and clean environment, especially in Barcelona and Boston. Paco, an active resident in the Casc Antic, shares his impression of feeling insulted by municipal planners and their decisions:

Of course, when they saw this zone, such a big hole, they said, "This cannot be a green area. How can we give these ugly people—who talk and do nothing—a green area?" . . . We were very moved to see how the municipality was taking advantage of these people and later how this neighborhood was abandoned.

In Barcelona immigrants confronted and still often confront discriminatory behavior every day in regards to unemployment and housing. Low-income Spanish residents—especially fragile elderly residents with small pensions—were particularly isolated and lacked resources for dealing with unhealthy housing, lack of public space, and daily filth. Many of them came from poorer regions of Spain in the 1960s and 1970s. In Catalan, people who arrived to Barcelona from Andalucia and Galicia were called *charnegos* (people who emigrated from a non-Catalan-speaking region) and often were scorned. Those who owned small traditional businesses in the neighborhood felt that they were slowly pushed away because their shops were not fashionable enough for a municipality or a business association that was eager to develop expensive shops in the neighborhood. They have felt sacrificed by a different vision for the neighborhood.

This context encouraged community leaders in Dudley, Casc Antic, and Cayo Hueso to help powerless residents fight against abandonment and marginalization. They fought for those who did not have the capacity to challenge the system or who feared that they would face retaliation for denouncing substandard living conditions and harmful environmental practices. In Dudley, people such as the Barros family organized residents to clean up trash in other families' backyards and find health providers for them. In the Casc Antic, as part of community resistance against neighborhood degradation, local leaders and workers helped older and vulnerable foreign residents resist expropriations and eviction and demand the renovation of degraded apartments. José, a coordinator of the community garden in the Casc Antic and an early community organizer, explains how public space maintenance work during the Forat fights was an occasion to discuss housing and sanitation needs with vulnerable residents: "[We organized assemblies] in the neighborhood to make sure we understood what was happening through the words of the people and to see in which ways we could confront the situations so that they did not see themselves as defenseless. Solidarity." In Havana, organizations such as Save the Children helped vulnerable groups that had social disadvantages and lived in unsecure housing structures.

Those commitments to the neighborhoods and their residents reflected a need to reevaluate the neighborhood as a place in the city where there should be equal treatment for all. Activists insisted on parks, playgrounds, and community gardens to be built not as temporary and second-rate spaces but from high-quality, sustainable, and sturdy materials. In Boston, Travis Watson from DSNI expresses the importance of equal treatment for all people:

It's just that sense that you are entitled to high-quality food being available, high-quality education, a decent living. You are entitled to having your street lights on all the time. Not just because you live in Brookline, you live in the South End, or North End. It's just because you are a human being. You are part of this community.

In Barcelona, a manifesto from community organizations that fought to revitalize the Forat asked for high-quality amenities and equal treatment for all residents. Signatories wanted the new gym to have "all the necessary provisions" and not be "a subterranean third-class sports center without proper ventilation and natural lighting."

Vulnerability and dignity are thus two social justice dimensions of activists' engagement, highlighting how historically, environmental justice and social justice are closely related. In Boston, Barcelona, and Havana, activists for social justice called for reform in urban planning decisions, housing policy, infrastructure development, and environmental protection practices. Their demands also reflected the evolution of the environmental justice agenda, through which a high quality of life for all people should be aimed for and achieved.

Combating Encroachment and Environmental Gentrification

Resisting Urban Development and Displacement

Questions of vulnerability, dignity, and discrimination emerge not only because residents experience racist and classist discourses and practices but also because broader urban transformations and processes have occurred in their neighborhoods. Many activists oppose municipal decisions and urban development practices in their neighborhoods, particularly in Barcelona and Boston where acute urban makeovers have taken place. The narratives of community activists in Barcelona's Casc Antic reveal that they contested broader urban policies in the city and rejected seeing their neighborhood and the whole city transformed into a theme park for tourists and high-income residents. The reconstruction of the Forat de la Vergonya (Hole of Shame) from an abandoned building site into a green area was a way for participants to express their disapproval of urban projects, encroachment, and tourism expansion in the Old Town. The green area was a symbol of their opposition to speculative private investment, as expressed by the 2002 Manifesto for a Green Old Town and without Urban Speculation: "We believe that a popular initiative to defend a green area from the threat of speculative projects and to humanize an urban space devastated by massive and random destruction deserves respect from public institutions."

Activists in the Casc Antic resisted a type of urbanism that ignores residents' daily problems and privileges showcase architecture projects. They were critical of the *Modelo Barcelona*—that is, the promotion of major events, the association of public interests with private money, and the central role of municipal planners in neighborhood rehabilitation. They feel that today Barcelona, and the old town in particular, is not a place to live in but rather a place to show, visit, work in, and consume. Local struggles challenged the public officials and planners who prioritized development in the city and neighborhoods and determined the roles that those neighborhoods play in the city for the profit of a few. To a similar extent, activists in Boston long distrusted the Boston Redevelopment Authority and its historic practices of endorsing the redevelopment of low-income communities. Nelson Merced, the former director of La Alianza Hispana, situated the early engagement of Dudley activists like him in the "real fear" of seeing houses in Dudley torn down and transformed into high-end condominiums for newcomers.

In other words, many community residents and their allies in Barcelona and Boston used their environmental engagement to fight municipal projects that promoted neighborhood renewal, land speculation, and gentrification. In some cases, city-led neighborhood revival and sustainability work actually becomes an environmental bad that affects community stability potentially more than locally unwanted land uses (LULUs), such as toxic sites. The racial aspect of whiteness is in some ways hidden and made invisible by the use of the words "green" and "sustainability" by local officials and decision makers. As recently rebuilt neighborhoods become desirable to new social groups, they push away existing residents. Newcomers move in attracted by the centrality and diversity of the neighborhood, but they contribute to erasing this diversity. As they "rediscover" a neighborhood, they displace a community that is very rooted to its place. In response to this process, many community workers—from organizations such as the Dudley Street Neighborhood Initiative (DSNI), Alternatives for Community and Environment (ACE), Project Right, and La Alianza Hispana in Dudley and Veïns en Defensa de la Barcelona Vella, the Associació de Veïns, and squatters in Casc Antic—started to fight displacement. They are afraid that new condominiums and populations would force residents to move to more marginalized areas of the city without benefiting from the improvements that they won after many long battles. The benefits of decades of protest and mobilization would vanish in just a few years.

In Barcelona, protesters saw gentrification as a product of a capitalistic system and a reflection of class struggles, which they aimed to address through their environmental engagement. According to them, young professionals move back to downtown Barcelona and displace older poor and immigrant populations through processes of environmental degradation and real estate reinvestment. The municipality also authorized changes in land use that allowed restaurants to open in areas that formerly had been zoned as residential. The words of long-term activist Paco reveal how his environmental work in the Forat served as a platform for fighting gentrification:

Everything else disappeared. And all the storefronts, as you can see, all are closed. And if we had allowed them to do this [create a parking lot], it would then be a park for snobbish people. The neighborhood got lost once already, and we did not want it to become lost again. And we believe that with a green area and with life there, then the neighborhood could resist this.

The green area that residents built in the Forat eliminated the possibility of private investment and speculation on the space.

Gentrification was also feared by Dudley activists, who were aware of traumatic urban renewal projects promoted by urban planners in other neighborhoods of Boston, such as the West End and the South End. For protesters, city officials often promised more beautiful neighborhoods—but only for a few. A community leader in Dudley called this process "Negro removal" and noted the gentrification that occurred in the South End (close to Dudley) in the 1980s and 1990s. As in Barcelona, many activists in Dudley participated in environmental projects as buffers to encroachment. Working collectively in an urban farm and organizing debates about the future of the neighborhood initiated long-term processes of resistance within Dudley. At The Food Project, program coordinators conducted youth leadership programs that taught young residents about changes in the neighborhood, including the national chain stores and hotels that opened outlets next to Dudley, and trained them to organize other residents to protest against these new investments.

In sum, the risks of environmental gentrification were high in Boston's Dudley and Barcelona's Casc Antic. When historically distressed communities become more livable, green, and welcoming, their ethnic and social composition tends to change in favor of whiter and wealthier groups (Checker 2011; Dooling 2009; Gould and Lewis 2012; Hagerman 2007). Brooklyn, Harlem, and the Bronx are often cited as examples of environmental gentrification. Today community organizations and NGOs are trying to balance working for environmental revitalization with avoiding the risks that residents will be displaced. Environmental justice advocacy can backfire, as happened in the Bronx with the organization We Act for Environmental Justice (WE ACT) (Checker 2011). Achieving revitalization without displacement is now a central dimension of environmental justice activism, especially in Boston and Barcelona. As Penn Loh, ACE's former director in Dudley, summarizes:

Around 2000, 2001, we were at a period of time where the economy was still very, very favorable to development. And so through the late 90s, early 2000s, we were seeing tremendous development pressure. . . . Broader economic forces were starting to look at the neighborhood again to do significant investment and reshape it without any input or any concern for who is living there now. They are just looking at it as, "This is cheap land that we can extract value out of."

Traditional "brown" concerns of the EJ movements pointed at policy makers and dominant institutions for treating marginalized groups as if they do not deserve to live in healthy neighborhoods and allowed the siting of waste sites where they lived. Green gentrification is the flipside of that process, through which the urban poor and people of color would only be allowed to live in less healthy and livable neighborhoods.

In Havana, real estate speculation and gentrification did not exist as they did in Boston and Barcelona. But the Talleres de Transformación Integral del Barrio (TTIB) (Workshops for the Comprehensive Transformation of the Neighborhood) were a response to government-sponsored urban renewal projects in the 1970s in Cayo Hueso. Through the TTIBs, members of the Grupo para el Desarrollo Integral de la Capital (GDIC) (Group for the Comprehensive Development of the Capital) responded to community needs and fought against the negative effects of urban renewal. Mario Coyula, a former urban planner within the GDIC, explains the problems created by the 1970s revitalization:

In the 1970s, Cayo Hueso experienced harmful redevelopment with [the construction of] two tall skyscrapers of twelve floors each. The idea was to sweep clean Cayo Hueso from Belazcoain to Infanta with two tall skyscrapers. The problem is that this led to the formation of empty spaces in the demolished areas. This is a model that was imported from Europe and France. There was no more street alignment. Those tall buildings became problems because there is a maintenance issue.

Through the TTIB, community workers tried to avoid community fragmentation and restore a sense of cohesion to the neighborhood.

Claiming a Right to the Neighborhood

As the risks of environmental gentrification and displacement became real, local protesters developed a narrative about a right to the city (Lefebvre 1968). Today, the right-to-the-city movement, particularly in the United States, connects groups that work on environmental justice issues, gentrification, and displacement so that they can address them in a mutually synergetic way (Steil and Connolly 2009; Harvey 2012; Mitchell 2003). The right to the city is held by the people who inhabit the city, not by those who own it. These inhabitants include the most marginalized and underpaid members of the urban working class and the most ethnically, culturally, and gendered alienated urban groups who demand collective rights on a space (Marcuse 2009a, 2009b, 2009c). Overlooking the spatial dimensions of urban inequalities makes it impossible to comprehend demands for justice and rights (Marcuse 2009a, 2009b, 2009c; Soja 2009). Beyond demanding a right to the city, in Boston, Barcelona, and Havana, activists also claimed a right to the neighborhood for all residents—to living in their home neighborhood and having it be a healthy environment.

Workers and community leaders in Dudley and Casc Antic have used the right to the city and the right to the neighborhood as they developed environmental projects. In the Casc Antic, social architect Hubertus Poppinhaus describes the meaning of his engagement:

A group of neighbors and I have been fighting for years to make this center more habitable, livable, for us, the people who live in it, who use it, and not only the people who do things for the people who pass by, tourists who use the city only in that sense. We live it more. . . . We want a city for the day to day and not only a city for partying.

Reestablishing residents' right to the city and neighborhood requires projects that reflect their daily use of the space and that minimize the presence of the municipality and capitalist interests in the neighborhood. For activists such as Barcelona lawyer Eduardo Moreno, the city should be built and evolve for its citizens and not the other way around. Activists present new environmental initiatives as commons that are enjoyed by everyone who helps build the neighborhood and who resides in it. Boston's Kathe McKenna from the Haley House Bakery Café created the business to bring healthy food—especially food that was not normally available in the neighborhood—to Dudley. As she says, "Now people can eat in their own community." They do not have to seek better food in other places. Food can be produced within the neighborhood and is not transported into other neighborhoods. Resources are not extracted from the place.

The right to the city and the right to the neighborhood also are reflected in activists' goals of spatially balancing environmental goods throughout the city and eliminating the environmental privileges of nonresidents. They often compare the environmental services and conditions that are found in high- and low-income neighborhoods and seek equally healthy conditions for low-income residents in their neighborhood. In Boston, the manager of the Leyland Street Community Garden compares the playgrounds and gardens in Dudley and other neighborhoods:

I drive through different neighborhoods, and I see what the city did for their park. They have nice playgrounds in different areas. But around here, they don't even come through here to even look at the places. There are some in Dorchester and Roxbury, Mattapan and Jamaica Plain, South End—all over the area around here. I go through all of them, and they all got beautiful gardens.

Her account identifies spatial discrepancies in the building and maintenance of community gardens, parks, and playgrounds throughout Boston.

Even in a more seemingly egalitarian context like Havana, people noted inequities among districts in the capital. Community leaders such as the artist Salvador from the Callejón de Hamel and Jaime from the Quiero a Mi Barrio gym noted that neighborhoods such as Miramar and Playa had many more green spaces and recreational facilities than Centro Habana, and they used those neighborhoods as reference points for what they aimed to achieve in Cayo Hueso. As Jaime says:

We say that we have the state for all the social domains. But Miramar has different conditions from us. Neighborhoods are not equal. In Cayo Hueso, there are people with a lot of needs. Many youth and their families are not able to pay to go to a place that charges them ten pesos to get in and to exercise and dance. Through the gym project, I want to give all people from all social spheres the ability to be here and enjoy themselves. Differences disappear in this project. Everyone has a place here.

Jaime's work addressed spatial inequalities in Havana, and he saw his gym is seen as as a symbol of overcoming unequal access to physical activity and sports in the city. He also noted that residents should not have to leave their neighborhood to use environmental resources in other parts of the city.

In sum, residents and their supporters in Boston and Barcelona and to some extent in Havana used their environmental revitalization endeavors as a platform for contesting broader urban processes, threats of encroachment, and spatial injustices affecting their neighborhood. They worked to reclaim the right of residents to the city and neighborhood by offering alternatives to the official discourses and practices promoted by local political and business figures.

Controlling the Land and Managing Borders

Reconquering the Land and Being Good Stewards

Activists in neighborhoods that are affected by encroachment, fragmentation, and gentrification commonly confront the back end of development—for instance, the health and environmental impacts of toxic waste sites and incinerators. But they also look at how land is used in the first place. They seek to transform land from a poorly utilized space to a true asset for local residents, who therefore must control the front end of development. Land is the site of struggles but also what is struggled over.

In Boston's Dudley, Barcelona's Casc Antic, and Havana's Cayo Hueso, residents and their supporters wanted to ensure that policy makers recognized the legitimacy of land uses and land tenure for community projects and also assert their own territorial sovereignty on the neighborhood. In the 1980s and 1990s in Boston, the Dudley Street Neighborhood Initiative (DSNI)'s advocacy efforts were centered on convincing the city of Boston that the open spaces created by residents were valuable and legitimate. A former environmental organizer in DSNI explains:

The people were very closely related to the land, particularly some of the older generation. . . . So much of our interaction around the city of Boston was the use of the land and coming to agreements that the use of the land for gardening was really a legitimate use. It's just as positive for the neighborhood as the development of a house.

In the neighborhood, the eminent domain status granted to DSNI in 1988 was a landmark victory, which led to the creation of a land trust called Dudley Neighbors, Inc. (DNI) for the area known as the Dudley Triangle. Since 1988, DNI has acquired vacant land and parcels owned by the city of Boston and leased them to private and nonprofit developers (for building affordable housing) and to individual homeowners and housing cooperatives. Because DNI can require that properties be used for purposes set forth by community members, residents now have a means of controlling development in Dudley. In addition to new homes, the Dudley Triangle has two community spaces, a community greenhouse, parks, and many community gardens, seven of them owned by the Boston Natural Areas Network through a land trust system.

In Cayo Hueso, several projects developed by the Talleres de Transformación Integral del Barrio (TTIB) (Workshops for the Comprehensive Transformation of the Neighborhood) have received a semipermanent form of land tenure. Among others, the TTIB worked to guarantee the permanence of a green area in the neighborhood through the creation of an urban farm, as Joel explains:

Rosita talked to the person in charge of urban agriculture, and she helped me a lot. We offered to have the urban farm on the land. This is now a way to guarantee a green zone. . . . We had a problem to solve and resorted to the urban agriculture office. It is now a green area, and there is a guarantee that they won't take it away.

Rosita also negotiated a change in land use for the area on which the Casa del Niño y de la Niña was built. Such achievements were significant in a country with central planning and whose citizens exerted little control over land and its development.

In Barcelona, the reconstruction of the Casc Antic divided activists, and many discussions took place between 2004 and 2006 to decide how the Forat de la Vergonya (Hole of Shame) was to be rebuilt and managed. Today the space is a permanent green area. Some activists (particularly squatters and older residents) wanted to keep the Forat and its surroundings as self-managed spaces, but others (mostly members of the Federation of Neighbors' Association and other more pragmatic residents) supported the city's decision to rebuild the area as a permanent green zone with municipal funds. In 2007, the Forat was remodeled into the Pou de la Figuera, and the municipality invested millions of euros in the surrounding area. Today many residents believe that this official rebuilding guarantees that this land will not be taken away for private investment and speculation.

Control over land also includes the creation and supervision of enclosed community spaces. In Dudley, the Haley House Bakery Café, founded by Kathe McKenna, provided a place that residents could appropriate for their own uses: "We are giving people a sense of access and control. The response from people who walk through the door is just that people feel so, so happy that this is there. It is a place that belongs to them. It's right in their neighborhood." In Havana, the Quiero a Mi Barrio gym allowed youth to build a feeling of ownership of a space that they appropriated through their daily sports practice. The founder built it for all of his neighbors, not as a personal and private business. Similarly, in Barcelona's Casc Antic, the sports association A. E. Cervantes–Casc Antic (AECCA) negotiated for new spaces for youth training at the Centre Esportiu and in local public schools. The sometimes tedious negotiations involved overcoming the resistance of local administrators to opening up schools after regular hours.

The environmental achievements obtained by activists in each neighborhood cannot be preserved without a strong stewardship system, and many community workers have enhanced residents' stewardship and involvement in new environmental projects. In Dudley, organizations such as the Urban Ecology Institute, the Boston Schoolyard Initiative, The Food Project, and the Youth Environmental Network have built strong connections among people, their land, and its resources to increase social capital around the territory and enhance the commitment that residents display toward the land. As a staff member from the Youth Environmental Network explains:

> This summer, we had youth employment programs that we helped support with twelve different organizations, and there was about 130 additional youth working in the park stewardship in community jobs. . . . And one is finding out about green space that they didn't know existed right there where they live. . . . And then community residents are walking by and thanking them, and I think that helps bolster their feelings. . . . So I think all of that helps to increase stewardship.

To a similar extent, the organization Earthworks involved residents in the maintenance and care of apple orchards and cultivated the image of orchards as precious commons that provide the community with a sense of permanence over time. If orchards do not receive proper care, they eventually become unproductive and die. Stewardship is particularly important for the flourishing of agricultural land, community gardens, and orchards.

In the Casc Antic, local organizers encouraged residents to be stewards of the new green spaces in the Pou de la Figuera by participating in the community garden and in environmental education workshops. Residents engaged in community policing to ensure that outsiders did not steal the harvest of the garden or destroy the playgrounds. In Cayo Hueso, the green-space clean-up brigades organized by Pablo and Rosa from the Casa del Niño y de la Niña enhanced residents' stewardship of environmental improvements in the neighborhood. Youth organizers taught children how to take care of new green assets and to share their accomplishment with their friends and families.

Protecting the land and being a good steward are crucial to maintaining new community assets so that they are not sold, abandoned, or allowed to become targets for illegal dumping. Community stewardship starts during the design process when residents reflect on the amenities that are needed to make it easier for them to take care of parks and community gardens and deter unwanted social behavior. Participating in the early stages of project development also allows residents to develop greater physical and mental ownership of a territory. Stewardship also must be associated with reinforcing parallel initiatives, such as building or housing maintenance by residents to make the neighborhood as a whole attractive and valuable for families.

Setting Up Borders

As local communities gain and retain protected access to land, they sometimes also set up borders against outsiders. Territorial control and the construction of borders often go hand in hand. In Cayo Hueso, Casc Antic, and Dudley, residents built borders to differentiate themselves from outsiders. At times, their efforts resonated with neighborhood sovereignty claims and demands for separation from outside groups. Borders (or boundaries) are physical restrictions that define tangible territories, serve as rallying points, and determine membership. They also can play the role of symbolic boundaries, which are "conceptual distinctions made by social actors to categorize objects, people, practices, and even time and space" (Lamont and Molnar 2002, 168). Boundaries are permeable, salient, durable, and visible and can be crossed, dissolved, activated, maintained, or transposed (Pachucki, Pendergrass, and Lamont 2007), so controlling them is important for community activists as they attempt to make the reconstructed neighborhood territory their own, make territory limits visible, and keep out undesirable outsiders. In the three neighborhoods in Boston, Barcelona, and Havana, some environmental projects served as borders that protected the neighborhoods against undesirable activities and groups.

In the early years of Dudley's organizing against illegal waste discarding, residents stood on street corners to denounce the companies that were dumping and to prevent their trucks from accessing empty lots. They also stood in front of contractors' trucks—using their bodies as borders between the dumping areas and the adjacent new community gardens. Spaces—such as the Dudley Town Common, which is a small neighborhood park at the corner of Blue Hill Avenue and Dudley Street—became borders themselves. Dudley Town Common marked a clear entrance to the neighborhood and showcased various community-built projects, such as the Jardín de la Amistad community garden. This landmark indicates where residents gained control in Dudley. In addition, fences, gates, and signs around community gardens created space markers that defined the ownership of the space, allowed residents to control the garden activities, and informed outsiders that they could not enter the space without permission. Gardens are spaces that need to be protected from outside threats, violence, and drug trafficking. As Nelson Merced, the former director of La Alianza Hispana, explains:

Developing gardens was basically an effort of not only controlling dumping but basically reusing the land in a way that discouraged dumping. Of course, we put a fence in so you couldn't get through. We didn't want to take any chances. And the other thing was that people wanted to plant. They didn't want somebody coming in and stealing their stuff in the night.

If the lots were left unproductive and unfenced, residents knew that there would be new waves of dumping. They also felt that green spaces that were built and maintained by residents and had a strong cultural identity might deter newcomers and help protect the space for residents.

In the Casc Antic, the first planting of a tree in the Forat in December 2000 was a symbolic and physical border against the municipality and its contractors, who were destroying buildings and leaving behind unsafe empty spaces. The tree was a shield that protected the lot against machines and bulldozers. In later years, the design, materials, and equipment used in the Forat and the Allada Vermell plaza constituted physical markers of difference between us (residents) and them (gentrifiers and tourists). Residents distinguished between the Born part of the neighborhood, which is full of *guiris* (foreigners and foreign tourists) and fashion and art stores, and Sant Pere and Santa Caterina, which are where residents live, work, and play. As residents redesigned the empty space of the Forat, they refused to build a *plaza dura* (a hard plaza made of concrete rather than of sand and grass). As Eva, a social architect, explains: "If in the Forat you let them lay down four bricks, then it will be a terrace as well. It is then a space that you deprive people of. What we really wanted was a green space, a green space, because if you give concessions and you cede, there end up being terraces."

The protesters wanted to redevelop the Forat and the Allada Vermell plaza with children's playgrounds and trees because they feared that outdoor restaurants and bars might take over the space and change its nature—making it into a consumption and party space. These fears were based on the previous conversions of some plazas in the Old Town into terraces that attract crowds of visitors who appropriate the space for themselves and privatize public spaces in the city. Furthermore, building a plaza that was a green soft space instead of a hard surface prevented cars from parking there and youth from other neighborhoods from skateboarding and creating noise. Soft surfaces with trees and plants also enhanced residents' spontaneous participation in the place. Neighbors could develop new pastimes together without the continued presence of police cars, which patrolled hard-surface plazas in Barcelona throughout the day.

Upon the rebuilding of the Casc Antic, community workers have helped residents reappropriate their revitalized Casc Antic territory and reconstruct a personal use of the space. As Silvana, a community worker from Iniciatives I Accions Socials i Culturals (INCITA), explains:

In the Forat, the girls appropriate the space for themselves. They use the Bicing [bike-sharing program]. These are girls with a veil, but they have a way to appropriate the resources of the neighborhood for themselves. I taught them how to use the bikes in the Parque de la Ciudadella. . . . It is a challenge because they feel a bit alien. There is a lot of exclusion. We want them to own the territory with activities like sports, music, and partying and to become the creative actors and residents of the territory.

In sum, activists promoted welcoming spaces for residents so that they could feel part of a collective endeavor but rejected border construction by the municipality, which was symbolized by the wall that police forces erected in the Forat on several occasions between 2001 and 2006.

In Cayo Hueso, several physical markers separated territories and uses of space in Centro Habana. The signs posted above two community centers (Quiero a Mi Barrio and Casa del Niño y de la Niña) point to spaces reserved for and dedicated to recreational activities for local children and young people. Areas such as the Casa del Niño y de la Niña and the urban farm were also fenced in to avoid dumping, theft, and destruction by outsiders. In the Callejón de Hamel, an iron and flowery entrance show where the more private space of Afro-Cubans starts and what types of activities take place inside the Callejón. Outsiders who do not approve of artistic environmental projects led by Afro-Cubans are discouraged from entering the space. The Callejón is a symbolic boundary against the dominant Cuban political views and culture. It contests what is culturally acceptable as artistic and urban revitalization projects. As Elias, its promoter, explains:

Salvador did this transformation and transgression from his own private initiative. He was a disturbance, and he was inspired by African cultures and religions. It was a symbiosis between the religion that brought slaves and the society that had to take them, and that was a slavery type of society. Some intellectuals don't want these words to be used.

The Callejón set up a clear dividing line between projects promoted by the government and one led by an artist like Salvador, which serves as a park and safe place to meet and play for residents, especially Afro-Cubans.

To conclude, residents and their allies in Barcelona, Boston, and Havana used environmental and health initiatives to resist outside threats and achieve control over neighborhood land and its borders. Yet, over the years many activists realized that they could not remain fully isolated despite the risk of undermining the cohesiveness and social fabric of their neighborhood. Today, they have to sustain a delicate equilibrium and frequently debate the need for outside support for the neighborhood. This external support highlights the tension between rejecting outsiders and needing selected technical and financial support. Over the years, activists have learned to debate, disagree, and build new local democratic practices in and for their neighborhood.

Changing Local Democratic Practices

Among others, exercising a right to the city involves rebuilding democracy at a local level. According to Mark Purcell (2008), local urban movements can contest free-market capitalism, foster the radical democratization and transformation of cities, and help to recapture its roots. As citizens take part in urban social life and in the political life and management of the city (Dikeç 2001), they can build different democratic attitudes (Purcell 2008) that contrast with representative and deliberative democracy, and the fact that it often neutralizes power. In Barcelona, Boston, and Havana, environmental revitalization projects have rebuilt communities and also have been the occasion for residents and their supporters to transgress existing norms for participation and civic engagement while questioning broader political arrangements and planning practices in the city.

Promoting Spontaneous Participation

Through their daily engagement in Barcelona, Boston, and Havana, community leaders created a new social world in the city and built a common framework through which new participants could become political activists. Spaces such as parks, urban gardens, and community centers were urban commons where activists promoted the collective management of shared resources. According to David Harvey (2012), building the commons depends on creatively appropriating already built public spaces, understanding common interests, and putting forth claims for common resources. Activists in the Dudley, Cayo Hueso, and Casc Antic neighborhoods of Boston, Havana, and Barcelona promoted self-determination practices as part of their new environmental projects and were skeptical about bureaucratic structures, traditional planning processes, and ideas proposed by urban elites. They believed that they had a right and responsibility to engage in a spontaneous, free, and even anarchical form of participation.

In Barcelona, residents and their allies joined a spontaneous movement to build a green area in the Forat, and it began without a formal organizational structure. Organizer Paco describes the spontaneity of the process:

We did not want to know anything, nothing about associations, about anything, and this came out in a people-based way. . . . We never wanted to be more than spontaneous neighbors. It was a space made by the neighbors for the neighbors where the presence of public authorities was not needed or desired to maintain, clean, and improve the space. Through the simple design of signs in the middle of the plaza, participants would just write their names down and explain what task they would be doing on a particular day—planting seeds, maintaining and cleaning up the garden, continuing the construction of the playground, etc.

The green plaza was a demonstration site for local participation that was independent and also well-functioning. Engaged residents emphasized "real participation" (*participación real*) as a nonhierarchical, almost anarchical intervention by people who were creative and inventive, who misused as few resources as possible, and who spent little or no money. According to Monika, a squatter: "We did not have a theory that we wanted to put into practice. It was more 'we like this,' and maybe it was very spontaneous and it grew over time." Residents often mentioned the work of a community leader who decided to plant tomatoes in the Forat and rallied others around him. Another example that activists recalled was the role played by Architects without Borders in gathering residents to design *charrettes* for the Pou de la Figuera.

In some cases, community leaders and workers in the Casc Antic promoted "self-management" (*auto-gestión*). For years, some groups that worked with youth autonomously managed the *campillo* (a sports and recreational area in the neighborhood) until municipal contractors demolished it in 2009. Recently, the Casc Antic residents won the right to self-management of the Pou de la Figuera space. Elisenda Ortega from the city of Barcelona values giving responsibility for the space to local people:

It is not the municipality that decides what goes there. No, in fact, there is an open calendar. People put their name in, propositions are being made, it's very dynamic. To have the keys means that you are the one responsible for the material, for the space, for a series of things—because it is a form of operation that is very close to self-management.

The Pou has been a learning space where residents take responsibility for the daily functioning of a new commons in the neighborhood with minimal municipal intervention.

In Cuba, physical improvements were made in the Cayo Hueso neighborhood that reflected the vision of the GDIC planning agency when it created the TTIB workshop—to promote community decision making. Over the years, residents identified priority areas, managed issues, and implemented solutions. As Mario Coyula, a former GDIC member, emphasizes:

A goal for the workshops was to promote horizontal initiatives and coordinate initiatives on the ground. The GDIC tried to promote the theme of participation—particularly conscious and active participation. These were our goals. In general, in Cuba, mobilization is very passive. It means carrying things from one place to the other or, traditionally, cutting sugarcane.

For instance, community workers acted to plan autonomous projects in Cayo Hueso, such as the creation of the Casa del Niño y de la Niña. Rosa looked for space where the Casa could be built, contacted families to win their support, and worked with children and the community to assemble materials and resources. As Arsenio Garcia, the main funder of the Casa, explains: "Rosa has a very special, spontaneous way of working and giving meaning to participation for kids, adolescents. She identified all the needs of the kids and the adolescents. She talked with them about what they wanted and how they would help." Over the years, despite governmental attempts at times to recentralize decision-making and planning processes, the TTIB workshop has been a model of citizen empowerment and decentralization. It transformed the way that people worked in neighborhoods and built local leadership and decision-making skills through participatory techniques.

During project development, some Cuban nonstate organizations helped teach community leaders how to manage community-grounded environmental revitalization without paternalism. For organizations such as the Martin Luther King Center, transmitting participatory practices to leaders was more important than the revitalization itself. The MLK Center used popular education tools to help residents contribute to project designs, seek and share resources, and make decisions. As its staff members trained community leaders, they strengthened civil and noninstitutional organizations in Havana so that spontaneous, innovative, and autonomous practices could emerge.

In addition, individual community leaders articulated a vision for a nonorganized form of participation in Cayo Hueso. The community gym Quiero a Mi Barrio emerged when neighbors joined Jaime to clean up some deteriorating buildings and rebuild them. People identified with the new space and started to contribute to its maintenance. For projects such as the Callejón de Hamel, participation was more transgressive. During the construction of the Callejón, the artist Salvador transformed an entire street and its neighboring buildings without the government's authorization. Only after structural environmental improvements became concrete and visible did Cuban authorities stop opposing his project.

In Dudley, much emphasis was given to a direct form of participation and a rejection of deliberative democracy. Residents experienced a new type of democracy through the projects in which they became involved, such as the construction of the parks and playgrounds for children. Tubal Padilla explains the broader meaning of his engagement: "[It is about] social usefulness by adopting a progressive agenda of a more direct democracy with the communities managing to control their own community and not only delegating power and authority."

Activists who constructed community gardens and parks felt that environmental projects help residents build healthier democratic practices and manage new commons. For instance, residents were involved in design meetings at all stages of the construction of the Kroc Center and the Dennis Street Park. The community was always the ultimate decision maker. Community gardens are also important tools for learning to discuss options, achieve consensus, and accept agreement or disagreement. As a staff member of the Boston Natural Areas Network explains:

When people come into the garden, the one agenda is gardening. So even people who don't speak the same language, I've seen communicate absolutely clearly over squash or tomato. You know, it's a profound community-building process, absolutely profound. Get a group of people to really work well together even if they don't like each other.

Creating a community garden implies that participants have built practices that lead to outcomes that satisfy participants and respect sound environmental management. People learn democratic processes and rules at the microlevel in their community gardens.

Deepening Democracy and Planning

Residents' attempts to create more direct and transgressive forms of democratic participation reflected their concerns about the municipality's neglect throughout the years and about what they considered as an undemocratic planning practice. In Dudley, residents resented the city's attempt to control their community project and also the elected officials' lack of representativeness. A community resident shares her anger: "You always have people from downtown or somewhere telling you what you need in your neighborhood." Community leaders used environmental projects to adopt a progressive political agenda that helped neighborhoods control their own development and not delegate authority to outsiders. They also called for a different planning practice and criticized the Boston Redevelopment Authority (BRA) for excluding residents of color from redevelopment processes, as happened in the 1980s. This mistrust of the BRA led organizations such as DSNI to conduct autonomous planning for Dudley, such as the 1987 comprehensive revitalization plan and the 1996 urban village visioning process. They established the idea that people of color should make decisions for themselves because are the ones experiencing the problems most directly.

To a similar extent, the struggles in Barcelona's Forat de la Vergonya (Hole of Shame) were evidence that the community was angry that the city and corporations such as the Promoción de Ciutat Vella SA (PROCIVE-SA) offered no meaningful opportunities for residents to participate in the redevelopment of the Casc Antic. Community activists were disappointed that planners historically did not consider their perspective, experience, and lay knowledge of their neighborhood and instead made decisions about redevelopment unilaterally without offering opportunities for debate. Virginia, an artist from the Companía Moki Moki, remembers her resistance against top-down planning practices:

I wanted to defend ordinary people. A lot of plans by the municipality exist for the economic revenue of Barcelona without thinking about the residents. It is only a question of parking and hotels. Plans get implemented without consultation. They think it's easy to cheat. They take people from the Forat out. We went to a demonstration because we don't like the comfort of accepting. You let them do one hotel, and they do twenty.

The Forat was rebuilt into a large green area, reflecting the activists' initial wishes for a greener neighborhood. Even so, many neighbors believed that the municipality hijacked the strength of the citizen's movement, invited only a few organizations to participate in the decision-making process (2005–2006), and misrepresented the Pou de la Figuera redevelopment plan as a consensus. As Rafael, a youth educator, explains: "The municipality has recycled the neighborhood actions into what they attempt to sponsor as their own project." Some community supporters called the practices of the municipality a "fantasmatic and fictitious type of participation led by people who are bought out." They believed that urban planners imposed the Planes Especiales de Reforma Interior (PERI) rehabilitation plans in the 1980s and forgot that during the dictatorship, they all worked clandestinely in the neighborhoods, committing themselves to serving the residents and their needs.

Protesters accused the municipality of undermining social organizations in the neighborhood by manipulating the concept of dialogue and not promoting real participation. These activists did not ask for a more inclusive and well-working deliberative or representative democracy but instead sought a more transgressive democracy. They also felt that public officials misused public resources when they reappropriated the neighbors' project since the plaza was already a green space built by residents and was really an attempt to impose municipal control over the Casc Antic. Several municipal workers themselves recall their own experience of how public agencies imposed their own plans on the residents:

It is true that in the area of public space, there was a very directing attitude by the municipality—that is to say, we were the ones who would offer what would be done and then it was done. At most, we would explain things to the neighbors. There was no bottom-up decision process, no debate, no consultation with the people affected by the changes.

As community leaders and residents organized to deepen existing local democratic spaces and planning processes in the city, their claims sometimes reflected democracy demands on the national scale. Activists in Cayo Hueso resisted the attempts of Cuban administrations to control neighborhood events by not giving community leaders the required paperwork on time. They became tired of having to give accounts back to public authorities and wanted some autonomy from the traditional rules and norms of the Cuban administration, which became stricter after the push for recentralization and vertical planning by the Castro regime in the late 2000s. As a GDIC member explains: "The workshop must be managed locally, but this is not permitted. The state is only sectoral. What is vertical is what rules here. It is a very strong structure with inertia." Serious disagreements erupted between community members (who wanted to transform social and political procedures and address the incapacity of the Popular Councils to respond to community needs) and members of the comfortable Cuban bureaucracy (who did not want to share their established power).

The local fights in the Casc Antic also were connected to Spain's national democracy, particularly for some older activists, such as Maria from the Associació de Veïns del Casc Antic, who became engaged in the neighborhood in the 1970s as a fight against Franco. They believed that after Franco died in 1975 after forty years of dictatorship, the old Franco loyalists imposed the monarchy of Juan Carlos on Spain without allowing citizens to have a say about their new type of nation and government. They believe that a real transition to democracy did not happen. This is why community leaders such as Maria and Pep took their national demands to the local level and brought them to the Casc Antic. Through their neighborhood activism, they hoped to achieve locally what they could not achieve nationally—that is, create a space for more radical self-determination, autonomy, and debate.

Discussion: Defending Broader Political Agendas as Conditions for Community Rebuilding and Place Remaking

Today, we know little about the broader political purposes and goals of historically excluded groups as they challenge their marginality, especially when they do so as part of long-term environmental revitalization. In the Dudley, Casc Antic, and Cayo Hueso neighborhoods of Boston, Barcelona, and Havana, complex political agendas—fighting racism and vulnerability, combating encroachment and environmental gentrification, controlling land and borders, and changing local democratic practices—developed at different stages of environmental projects, strengthened community rebuilding and place remaking, and ensured the long-term sustainability of projects. The new environmental spaces that activists built are repositories of broader political goals, values, and principles. As they worked on new community gardens, playgrounds, and green spaces, residents and their allies resisted the disruption and degradation of their neighborhood as well as the violence of private accumulation, as illustrated by activists' critique of broader urban developments such as encroachment, speculation, and tourism. Their fight was anchored locally but also had a citywide projection and dimension. Urban developments and change were political targets that are at the foreground of neighborhood fights. Urban developments ultimately represented the socioeconomic and political dynamics that marginalized residents had to confront to address environmental and health injustices and help preserve new environmental commons over time.

In this chapter, the fights of activists shed new light on the right-to-the-city concept as residents fought against the privatization of public spaces and developed new uses for spaces according to their own vision. The framework of the right to the city is also a right to the neighborhood. For residents in Cayo Hueso, Dudley, and Casc Antic, guaranteeing a right to the neighborhood involved fighting the stigmas that were associated with their place while restoring a sense of dignity to residents. Activists wanted to change the connotations that the media, local politicians, and even residents from other neighborhoods assigned to their place and increase its legitimacy in the city. To counteract territorial stigmatization and the social harms that stigmatization had on residents (Wacquant 2007), residents and their supporters addressed stereotypes about low-income and minority residents. Resistance also involved helping residents who faced eviction, were victims of housing abuses, and were treated disrespectfully by landlords, private developers, or municipal planners. The greatest challenge for the three neighborhoods was to achieve environmental revitalization for residents without indirectly fostering processes that might directly or indirectly exclude them from the neighborhood. Struggles for affordable healthy housing and against environmental gentrification are ultimately meant to guarantee sustainable environmental justice as residents are provided with the means and support to remain in the place they helped rebuild.

The stories of environmental revitalization in Dudley, Cayo Hueso, and Casc Antic reveal the centrality of space in local mobilizations. Environmental mobilizations are situated and rooted in a specific site. Struggles are linked to notions of spatial equity and attempts to provide environmental goods as a way to address spatial inequalities. Space is a constitutive element of collective action and is not merely in the background. Neighborhoods are iterative sites of mobilization and an emblematic place of contention for the struggle of poor communities against spatial inequalities. Thus, resistance to outside forces and development of community-based projects require control over the land of the neighborhood. Residents expressed a strong connection to the parks, recreational areas, community gardens, and urban farms in their neighborhoods. Their vision was to control the long-term projects and activities that took place on this land, gain tenure over it, and ensure its stewardship. These goals require clear markers that delimit their territory and discourage outside forces (such as private developers, city officials, police forces, and gentrifiers) from entering or using it. Parks and community gardens and the activities organized around them serve as buffers against outsiders. Porous boundaries are also present internally within the neighborhood because some environmental spaces (especially in Boston and Barcelona with community gardens) are open only to members and only certain groups are welcome in the new spaces.

Finally, environmental initiatives are commons that need to be preserved and learning spaces that can strengthen a new type of democracy and planning practice in the neighborhood and the city. Indeed, the right to the city also involves the reconstruction of a new form of democracy (Purcell 2008). Building the type of neighborhood that community members envision cannot occur without asking who makes decisions in the neighborhood, whom they benefit, and which benefits are distributed. Environmental projects thus become a tool for questioning broader political arrangements in the city, empowering residents, and creating new, spontaneous, self-managed, and at times anarchical forms of participation and decision making. In all three neighborhoods, municipal decision makers and planners imposed their projects on every piece of land in the neighborhoods and often left residents out of the process. As a counterpoint, residents, community leaders, and neighborhood workers offered through their environmental endeavors a space for debate that formerly did not exist. Some of their goals was to promote a more direct form of democracy rather than a deliberative democracy that they felt does not lead to the meaningful participation of community members and only reinforces power imbalances. They thus wanted to transgress and deepen existing democratic practices in the city and at times in the country as a whole, reflecting nation-building claims.

The stories of activists reveal that many of them could not live without being engaged in political protests and contestation. Their lives became focused on political organizational work in their neighborhood, which at times led to conflicts and instability—between radical activists and activists who were willing to negotiate with the municipality. In each place, some activists were more extreme than others. Some (mostly older leaders and squatters in Barcelona, more radical community leaders in Cayo Hueso, and long-time residents in Dudley) refused to engage with authorities and distrusted all government representatives, while more pragmatic leaders and community workers were more willing to enter in dialogue with municipal officials and planners. Neighborhood cohesiveness and organization as well as the effectiveness of collective mobilization were thus at stake, but such internal debates also strengthened local democratic practice and learning.

Although all three neighborhoods had some commonalities in political agendas, there were real differences among neighborhoods. In Cayo Hueso, urban developments promoted by private interests and supported by public officials were not nearly as harmful as they were in Dudley and Casc Antic. Despite the increasing level of foreign tourism investment in Cuba, its influence is limited but growing, with the recent opening of places such as the Palacio de la Rumba, a new venue that was planned to attract tourists to rumba events. Threats of encroachment and gentrification were more immediate and relevant in Barcelona and Boston. In Cayo Hueso, activists fought to defend their space against omnipresent and controlling public authorities as well as against stigmas about the neighborhood and its Afro-Cuban residents. They attempted to control their land and its boundaries to ensure that they had—for the first time since the Cuban revolution—an autonomous and spontaneous right to participate and decide issues in their neighborhood. In an autocratic regime such as Cuba, this right could not be as open and visible as it was in Boston and Barcelona, and activists had to be more strategic in how they claimed a right to their neighborhood and attempted to control the activities in it.

Activists in Dudley and Casc Antic and less so in Cayo Hueso desired to achieve greater protection with some self-segregation. As community leaders and their supporters attempted to control the land and its borders and provide a sense of protection for residents, they also constructed a self-sustained and contained community with a select membership. In some ways, they reprivatized public spaces for their own uses. They appropriated the territory for themselves and made it exclusive, and this occurred with tacit approval from the municipality. Living in a socially and racially homogeneous neighborhood reflected a desire to protect oneself against threats to the fabric of the neighborhood, which can cause feelings of loss and alienation (Hummon 1992; Brown and Perkins 1992). In Barcelona, homogeneous spaces strengthened racial, socioeconomic, and political identity. They brought together low-income Spanish residents with immigrant populations, squatters, and intellectuals. More racially and socially homogeneous spaces provided a space for identity formation and confirmation. They also soothed residents who suffered from the pressures of interethnic relations and bolstered their ability to deal with negative racial relations (May 2001). In all three cities, some of the activists' preference for a more homogeneous neighborhood also stemmed from a desire to protect residents from discriminatory policies and attitudes and to empower them through training, leadership development, and political activism activities. However, this goal (protection; control over group membership and space use) can become problematic when it clashes with another goal (just cities that are diverse and welcoming for all). How can vulnerable groups be protected without creating new forms of spatial exclusion?

7

Conclusion: Toward a New Framework for Place-Based Urban Environmental Justice and Community Health

On a sunny March afternoon in the Dudley Greenhouse, Jennie Msall and Danielle Andrews from The Food Project marvel at the crops grown by refugees from the Boston Center for Refugee Health and Human Rights. Throughout the winter, gardeners have tended their crops while discussing their new lives and resettlement in Boston with one another and with their clinician. For refugees, the greenhouse is a space for growing plants, processing and healing trauma, and learning about living and thriving in a new city away from the defined boundaries and schedules of hospital settings. For local young people, the Academic Year Program (AYP) at The Food Project addresses food-access issues in Dudley, build raised-bed gardens in high schools, teaches technical skills (about farming, sharing, and selling food), develops leadership and public speaking skills, and builds positive objectives for their neighborhood and families. Through nonprofit groups like The Food Project, Alternatives for Community and Environment (ACE), and the Dudley Street Neighborhood Initiative (DSNI), neighborhoods such as Dudley have been transformed from blight to revival. Environmental projects in cities provide a medium for socializing but also help repair traumatized and fragmented communities and remake places for residents.

This book examines how active residents, leaders, community workers, and their supporters in three emblematic and marginalized urban neighborhoods fought for greater environmental quality and livability. In Dudley in Boston, Casc Antic in Barcelona, and Cayo Hueso in Havana, neighborhood activists enhanced the environmental revitalization of their neighborhoods while helping residents to stay. They supported successful projects, including urban farms, community gardens, farmers' markets, parks, playgrounds, sports centers, community centers, and waste-management initiatives. They defied the odds against them by gathering support beyond the neighborhood from nongovernmental organizations and from city funders and some planners. Activists used creative collage techniques and joined with broad and often unexpected coalitions of sub-community networks to jumpstart their projects. They addressed negative stigmas and images about the neighborhood and brought pride and hope back to residents.

These observations across neighborhoods raise simple questions: How do similar patterns of mobilization for environmental revitalization arise in cities and contexts that do not resemble each other at first glance? What motivates residents to take action in their neighborhood and remain committed to it over time despite the dire baseline conditions of the place and the multiple obstacles to transforming it? How do physical interventions (such as parks, playgrounds, community gardens, and healthy housing) connect to underlying goals? How do issues of health and health quality play out in local struggles? And how do people's experience of and feelings for their neighborhood shape their engagement in environmental revitalization? In this book, I examined how activists used physical environmental and health improvements to reshape their relationship to their place and their relationships to the political institutions, processes, and actors that affect it. I also placed activists' struggles within broader historical processes of uneven development, urban growth, valuation and devaluation, and degradation to understand their engagement and its meanings and values, as well as the broader political purposes and goals that historically marginalized groups frame as they make sense, resist, and challenge their marginality.

Neighborhood Transformation in Boston, Barcelona, and Havana

The Closed Circle of Environmental Justice: From Broken Neighborhoods to Holistic Environmental Revitalization and Community Rebuilding

The transformations in the Dudley, Casc Antic, and Cayo Hueso neighborhoods of Boston, Barcelona, and Havana in recent decades are evidence that the livability and environmental quality of distressed and marginalized neighborhoods can be enhanced through a comprehensive revitalization of places and spaces. Community leaders, local workers, and active residents acted in complementary domains that triggered positive snowball effects over time and reflected a natural continuum in their minds. Environmental initiatives were more holistic than they traditionally are presented because activists did not envision their work in isolated compartments or sectors. As marginalized communities fight toward environmental justice in cities, they do not stop at struggles against clearly identifiable brown contamination sources but go much further. Activists anchored projects in one area of environmental justice as a stepping stone toward related environmental initiatives.

Starting with land clean-up and waste management, residents and their supporters then redeveloped land by developing community gardens and producing healthy food, enhancing open spaces, and creating green areas. They also worked to provide sports and recreational areas for youth and families that at times can be used for education and learning programs. Finally, residents' habitat quality improved through advocacy for healthy and affordable housing and for economic security in the neighborhood. Revitalization projects played a comprehensive role in transforming neighborhood conditions and increasing residents' quality of life. Such endeavors represent what I call the closed circle of environmental justice—from waste sites and empty wasted land to safe and affordable food, outdoor public spaces, indoor community centers (for sports, play, and learning), healthy housing, neighborhood affordability, and economic security.

Over the years, community revitalization efforts in these three neighborhoods were successful thanks to the large, diverse, and often unexpected coalitions of residents, local organizations, architects, artists, funders, political leaders, and at times environmental NGOs that mobilized for the neighborhood. These diverse coalitions worked on many interrelated aspects of long-term environmental justice. Coalitions of supporters offered residents financial, technical, and political help as well as material and nonmaterial resources, which they weaved together through bottom-to-bottom networks. For activists, the neighborhood landscape was both a material and symbolic resource, and they took advantage of sociospatial dynamics and strong spatial capital within each neighborhood. They displayed similar interests in neighborhood reconstruction and community power and often shared common values of solidarity, altruism, sharing, and defense of vulnerable residents. At times, partnerships were loose but tightened during crucial moments of neighborhood organization when common interests in specific projects took precedence over differences in organizing traditions and long-term agendas.

Unlike what previous research has claimed (e.g., Hunter and Staggenborg 1986), foundations such as the Kroc Foundation and the Riley Foundation (in Dudley) and international NGOs such as Oxfam (in Cayo Hueso) did not encourage community groups to abandon more radical claims. In fact, foundations and international aid agencies felt as though they were part of the local community and at times activists themselves. They were rooted in the neighborhood and its daily struggles. In Dudley, Casc Antic, and Cayo Hueso, there were three types of activism: street activists (such as community leaders and volunteers) wove together material and nonmaterial resources to organize protests and complete environmental projects; technical activists (such as environmental NGOs, social workers, universities, architects, and planning groups) provided technical, legal, and scientific expertise and at times advocated to different policy makers on behalf of residents; and funder activists (such as Community Development Coporations, foundations, and aid agencies) provided financial support for projects, who also at times lobbied the city on behalf of neighborhoods groups.

These three activisms intervened at different moments of project development and at times combined their work to use and even shape a political context (national laws, municipal policies, policy makers, and funding resources) that was receptive to their needs. At the local level, national policies and constraints (such as living in an autocratic regime like Cuba) did not matter as much to activism development because residents and their supporters had leeway to develop their work. Political authorities were also interested in fostering environmental improvements for vulnerable neighborhoods and were willing to accommodate innovative arrangements in the city. Sometimes their willingness to accept innovation was because political stability was sought (in Cuba) and because elections were upcoming (in Boston and Barcelona).

Remaking Place, Fighting Trauma, and Strengthening Identity
The neighborhood is an important place for marginalized groups (Clark 1989; Manzo 2003; McAuley 1998; Pattillo 2007; Falk 2004). People's attachment to their place of residence and to their community is strongly tied to the formation of their identity and the protection of this identity (Altman, Low, and Maretzki 1992; Low and Lawrence-Zúñiga 2003; Kefalas 2003). Place identity originates in people's relationship to the physical, political, and environmental world around them and is also shaped by their experiences and interactions with others (Fredrickson and Anderson 1999; Proshansky, Fabian, and Kaminoff 1983; Bondi 1993; Scannell and Gifford 2010). In distressed neighborhoods, residents have created particularly strong roots and developed bonds of mutual support (Manzo, Kleit, and Couch 2008). Place attachment often motivates residents to interact with neighbors and invest time in it (for instance, by monitoring developments in the neighborhood or participating in community

planning) (Chavis and Wandersman 1990; Cohrun 1994; Davidson and Cotter 1993). It provides a sense of security and well-being for residents, defines boundaries between groups, and anchors memories, especially against the passage of time (Gieryn 2000; Logan and Molotch 1987). The stories of activists in Cayo Hueso, Casc Antic, and Dudley reveal that place attachment and sense of community were motivators for local action (Chavis and Wandersman 1990; Cohrun 1994; Davidson and Cotter 1993). That said, residents did not fight against traditional threats to safety, property, or social programs or from contamination (Blum and Kingston 1984; Cox 1982; Davis 1991; Fainstein 2006; Fisher 1984; Gotham 1999; Pattillo 2007; Peterman 2000; Venkatesh 2000) but instead worked for long-term improved environmental quality and health outcomes. Residents were proactive. In Dudley, Casc Antic, and Cayo Hueso, activists engaged in environmental revitalization projects inspired by an emotional connection, a strong sense of place, and an attachment to their neighborhood. They felt responsible for the well-being of their community and experienced personal growth when they helped other residents by transforming the neighborhood.

Community connections, place attachment, and positive emotions toward the neighborhood led residents and their supporters to engage in local environmental projects, such as lot clean-up, park construction, and urban farming. Activists sought to remediate the neighborhood's loss of identity and dismantling of its social fabric. The loss of place has indeed devastating consequences for individual and collective memory, identity, and mental wellness (Fullilove 1996), and activists' projects addressed years of direct and indirect destruction and abandonment, a sense of being in an urban war, and environmental violence and environmental trauma. Environmental projects were meant to repair a broken community and heal both individual (Gerlach-Spriggs, Kaufman, and Warner 2004) and collective wounds. They support the environmental recovery of sites but also groups of people and help to overcome two types of negative experiences—those of long-time residents who witnessed the decay of their neighborhood and its environment and those of new residents who were uprooted from their homeland or community and whose resettlement into a new (and fragile) community was accompanied by loss and hardship.

Environmental justice and safety are also closely connected. Many new environmental spaces are safe havens that offer soothing, healing, protection, and nurturing. They are physical and emotional spaces where marginalized residents can reinhabit neighborhood territory, become visible again, reclaim memories, and redraw positive connections to their place. As people rebuild their neighborhoods, many rebuild themselves. This was especially important for African American and Latino residents in Dudley, Afro-Cubans in Cayo Hueso, and Spanish migrants and Latin American and African immigrants in Casc Antic—people who often were invisible or stigmatized in the city but can now treasure these new neighborhood spaces.

More broadly, place remaking can serve to construct a self-contained urban village and promote community flourishing. Activists in Dudley, Cayo Hueso, and Casc Antic used their environmental initiatives to build an urban village with a tight social fabric, self-sustainability, traditional farming and recreational practices for migrants, opportunities for socializing, protection, and transmission of customs and knowledge. In parks and community gardens, traditions such as community work were revived and passed on to younger generations. In addition, community workers used environmental projects to appease tensions, address preconceived opinions, and promote colearning and sharing between different ethnic and cultural groups. Environmental endeavors were thus oriented toward the inside of the neighborhood and toward remaking a Gemeinschaft where residents collaborated, showed altruistic behavior, and were bound by cultural ties and values.

Finally, environmental improvements and engagement in local projects strengthen residents' identity. Place remaking is a dynamic and dialectic relationship with a positive feedback loop. Local identity is reshaped and remade through the recreation of a positive memory during project development, with an emphasis on celebrating the community with all its diversity, history, and practices. Multicultural festivals in community gardens, murals in rebuilt public spaces, and artistic pieces in new parks help residents to construct a new positive identity and indirectly encourage participants to continue to participate in reconstruction projects. Local leaders and workers thus focus on two delicate aspects at the same time— promoting neighborhood diversity and enhancing unity for ongoing and future fights. Their engagement demonstrates that contemporary global cities do not have to lose their shared histories, adopt only transnational identities, or privatize all public space.

In short, place played a dual role in Dudley, Casc Antic, and Cayo Hueso as both a motivator for action and also a goal to be achieved. It also had both negative connotations (of loss, suffering and grief) and positive ones (fond memories and hope), which shaped residents' relationship to their neighborhood. Environmental justice initiatives in vulnerable urban neighborhoods help residents remake their place, make their neighborhood flourish, and strengthen identities.

Visions for Broader Political Changes in the City
In addition to projects that were oriented to the inside of the community, activists in these three neighborhoods also framed initiatives in relation to broader urban changes and political goals. In Barcelona, Boston, and Havana, activists in the Casc Antic, Cayo Hueso, and Dudley neighborhoods realized that the decay and losses in their neighborhood required mobilization to revitalize them. They framed their work not as separate from the larger city but as part of political and socioeconomic processes that affected neighborhood stability, dynamics, and cohesion. Activists advanced political goals that confronted urban socioeconomic and political pressures, addressed environmental and health injustices, and allowed fragile residents to remain in their neighborhood.

The transformation of the urban economy toward decentralization, globalization, technology, finance, and services has been accompanied by rising socioeconomic inequality (Friedmann 1986; O'Connor 2001; Sassen 1998). Urban movements have resisted the disruptions, degradations, and privatization of their neighborhoods (Harvey 1981; Smith 1982). Recently, movements have coalesced to build a revitalized, cosmopolitan, just, and democratic city (Fainstein 1999, 2006). Such values are often represented in the right-to-the-city movement, which states that the people who inhabit the city, not those who own it, hold that right. Often, the most marginalized and underpaid members of the urban working class and the most ethnically, culturally, and gendered alienated urban groups demand collective rights (Marcuse 2009a, 2009b, 2009c). A vision for economic and environmental justice is often combined with protests against real estate speculation, privatization of public space, and gentrification and struggles to secure rights to land, democracy, and human rights (Steil and Connolly 2009; Marcuse 2009a, 2009b, 2009c; Mitchell 2003).

In Dudley, Casc Antic, and Cayo Hueso, local community leaders, active residents, and neighborhood organizations used the right to the city as their basis for challenging the public officials, planners, and media that controlled the developments to be prioritized in each neighborhood and the neighborhood's "place" in Boston, Barcelona, and Havana. Prejudicial beliefs were attached to each neighborhood, which in turn limited the creation of new opportunities and reinforced residents' marginality (Wacquant 2007; Garbin and Millington 2012). Through new environmental initiatives, activists fought existing racist and classist stigmas and stereo-types about low-income and minority residents—that they do not care about the long-term environmental quality of their place, that they ne-glect their health, and that their practices are not socially acceptable.

Protesters also linked traditional right-to-the-city demands with ele-ments that include a right to the neighborhood—a right for residents to live in it, use it, and enjoy high environmental standards in it. They pushed for new environmental goods so that individuals and families did not have to seek these goods and spaces in other neighborhoods. Spatial equity and fair distribution of environmental goods were central to their fight. Many community leaders and neighborhood organizations addressed abuses that were caused by municipal decision makers and developers, including situations of extreme precariousness, such as mob-bing (pressuring apartment dwellers to vacate their apartments) and evic-tions in Barcelona. They brought about conditions that allowed residents to claim their dignity, believe that they have a legitimate place in the neighborhood, and address environmental privileges that other neighbor-hoods have.

Most important, community organizations, NGOs, and funders tried to maintain a delicate balance between achieving environmental revital-ization and increasing risks of displacement for fragile residents. They attempted to provide greater environmental goods for residents without indirectly inviting people with higher purchasing power, high-end stores, and eventually increased real estate prices to the neighborhood. When blighted communities become more livable and green, their ethnic and social composition tends to change (Checker 2011; Dooling 2009; Gould and Lewis 2012; Hagerman 2007). Minority and low-income residents in Dudley and Casc Antic were at risk of being directly or indirectly priced out of the neighborhood and seeing waves of newcomers replace them, which would mean that their activism backfired on them. Even in Cuba, neighborhoods such as Cayo Hueso are now changing as new concert halls such as the Palacio de la Rumba are reshaping the image of the neighborhood. The Palacio is a place for tourists and others who can pay in hard currency. Ironically, new environmental goods and services in Barcelona, Boston, and Havana might affect the cohesion and well-being of the neighborhood and its vulnerable residents far more than toxic waste sites and blight did. Achieving revitalization without gen-trification and displacement is thus a crucial dimension of environmen-tal justice. Struggles against gentrification and for an affordable healthy habitat ultimately can help achieve sustainable environmental justice in each neighborhood as residents have the means to remain in the place that they helped rebuild.

Consequently, resisting external pressures and developments needs to go hand in hand with controlling land and territory and achieving neighborhood sovereignty. In each neighborhood, residents built strong connections with the land and its uses and local activists wanted to ensure that parks, gardens, and recreational areas were not transformed into private developments such as expensive condominiums and restaurants. To exercise secure tenure and stewardship over the land, residents need to participate in communal work, park and playground maintenance, and community policing over farms and gardens, which concurrently leads to a stronger sense of ownership over the land. Land control also translates into the construction of physical and symbolic borders that delimit the neighborhood territory from a dominant society and its practices. Parks, gardens, and green spaces—such as the Jardin de la Amistad in Dudley, the Forat de la Vergonya (Hole of Shame) green spaces in the Casc Antic, and the Callejón de Hamel in Cayo Hueso—are borders that activists created to delimit the neighborhood and its members.

Finally, environmental initiatives are commons that serve as learning spaces and allow a new type of democracy and planning practice to emerge in each city. Building the type of neighborhood that community members envision cannot take place without asking who makes decisions in the neighborhood, who benefits from those decisions, and what are the benefits themselves. Activists challenge broader political arrangements in the city as well as existing democratic planning and participation practices. They are recapturing the roots of democracy (Purcell 2008). In all three neighborhoods, municipal decision makers and planners originally imposed their orders on every piece of neighborhood land without allowing contributions from the residents. Residents now have a space for debate that did not previously exist in the city. As they developed environmental and health initiatives, they created self-managed commons that functioned as experiments for a new form of democratic planning and participation. They promoted a direct form of democracy rather than a deliberative democracy, which they felt is bound by power imbalances and did not lead to meaningful community participation in neighborhood decision making. Their demands were more transgressive than those traditionally presented by theorists in the just-city movement: they asked for self-management of land and projects and for more spontaneous and anarchical forms of participation and dialogue. Many activists saw themselves as part of a political trajectory and seemed at times to be in self-representation mode. They still can not live outside a life of political claims, protests, and contestation.

In sum, urban marginalized neighborhoods from these three countries share many similarities. In all three neighborhoods, activists worked to build more just, healthy, and vibrant communities, and they shared similar processes and outcomes. In Boston, Havana, and Barcelona, residents and leaders of distressed neighborhoods constructed similar forms of attachment, experienced longtime exclusions and traumas, and developed a sense of responsibility and care for their neighborhood, which led them to focus on comparable actions, agendas, and strategies in their revitalization projects. Similarities (in experiences and objectives) are more important than differences (in development settings and political contexts) in accounting for holistic environmental revitalization work, community reconstruction, place rebuilding, and the defense of broader political agendas. At the local level, differences in levels of development and political systems do not matter very much to the development of community-sponsored initiatives as individuals and groups tend make sense of their neighborhood's exclusion from the rest of the city and of periods of decay and abandonment in a similar way. Place attachment and a strong sense of community also motivates them to act and build support around them.

Differences between Neighborhoods and Internal Tensions

These commons patterns of neighborhood engagement do not mean that the three neighborhoods had no differences or nuances. The baseline conditions in each place revealed inadequate housing conditions, substandard waste management, and inadequate green spaces and recreational facilities. However, nuances exist within each place. In Cayo Hueso and Dudley, residents suffered from malnutrition because fresh produce and healthy food were not always readily available. Dudley residents had high rates of obesity and cardiovascular diseases, and Cayo Hueso residents were often undernourished, which was particularly alarming during the first decade of the Special Period after the collapse of the Soviet Union. In the Casc Antic, farmers' markets and fresh produce were found throughout the neighborhood, although residents complained that gentrification drove up food prices, and groups such as the Xarxa de Consum Solidari worked to increase access to healthy and affordable local food.

Neighborhood abandonment and decay took different forms in each neighborhood. In Dudley, residents lived for two decades amid violence, arson, and dumping that the authorities ignored. By the mid-1980s, 1,600 vacant lots made the neighborhood look like a war zone. In the Casc

Antic, public authorities attempted to revitalize the old town after the return of democracy in 1975, but the projects they promoted resulted in negative social and environmental effects, such as mobbing (forced relocation), evictions, dumping on vacant lots, and abandonment of buildings. Finally, in Havana after Castro's 1959 revolution, planners neglected old neighborhoods such as Cayo Hueso and allowed buildings to become structurally unsafe and public spaces undesirable and unwelcoming.

Some variations existed in experiences of racial discrimination and social exclusion. First, racial discrimination by authorities, developers, and other residents was more violent in Boston than in Cuba or Barcelona. In Cuba, racism against Afro-Cubans was manifested in private spheres, particularly against Afro-Cubans' social and cultural traditions, despite government policies to eradicate racist attitudes since the beginning of the Cuban revolution. In the Casc Antic, residents faced discrimination regarding their social origins, jobs, and stores. Older residents from rural Spain who moved to Barcelona in the 1960s, 1970s, and 1980s felt that there was and is no place for them in a redeveloped fashionable old town and that they were being excluded from the remodeling of Barcelona. Latino and African immigrants in the Casc Antic felt and still feel harassed by derogatory attitudes toward them and a heightened police presence in the neighborhood. Finally, Dudley residents lived with unwanted social behaviors such as crime and violence, which added a layer of difficulty to their environmental revitalization work because empty lots and abandoned green areas were often occupied by drug dealers and shootings were frequent. In the Casc Antic and Cayo Hueso, crime and drug dealing were more limited.

To respond to the substandard conditions in their neighborhoods and the planning processes that marginalized them, residents began a series of environmental projects, with each neighborhood taking a slightly different path. In Dudley, activists started by working on massive clean-up campaigns, protesting against illegal waste-management practices, and turning empty lots into parks, playgrounds, and community gardens. Such transformation was made easier after DSNI gained the power of eminent domain over the Dudley Triangle. Other neighborhood organizations and leaders in Dudley helped to build technical skills and promoted physical activity and healthy eating. In the Casc Antic, residents fought for six years against the city of Barcelona over the redevelopment of large empty lots in the Forat section of the neighborhood. As activists resisted municipal plans, they converted the Forat into large green spaces, playgrounds, sports grounds, and community gardens. After years of conflict, the municipality in 2007 finally agreed to build a green zone. Since the 2000s, residents also have successfully advocated for improvements in sanitation, building structures, waste management, green space, and opportunities for physical activity. In Cayo Hueso, residents benefited from Castro's support in the late 1980s for more autonomous participation within the TTIB workshops, and they facilitated structural improvements to buildings, enhance waste management, and urban agriculture. Other leaders worked independently on improving opportunities for physical activity for youth, environmental capacity building, and art and cultural projects.

Although there were similarities in political agendas among activities in the three neighborhoods, there also were differences among neighborhoods. Cayo Hueso did not have as many private urban developments sponsored by public officials as Dudley and Casc Antic had. Despite foreign investments in clubs, restaurants, and hotels in Havana, until recently such investments did not have a real presence in Cayo Hueso, where threats of encroachment and gentrification were not as pervasive as they were in the Casc Antic and Dudley. In Cayo Hueso, activists tended to defend their space against omnipresent and controlling public authorities as well as against stigmas about the neighborhood, especially about Afro-Cubans. For the first time since 1959, activists carved out an autonomous and spontaneous right to make decisions about their neighborhood and its boundaries. In an autocratic regime such as Cuba, such a right could not be expressed as openly as it was in Boston and Barcelona, but even so, activists were vehement about their right to self-manage green spaces and projects in Caya Hueso. Their distrust of public authorities led many of them to refuse to negotiate with officials. In Dudley and Boston, activists sought greater protection and even isolation from gentrifiers, tourists, and encroachers of various other kinds. Residents feared that their neighborhood would be taken over by outsiders and that poor families and residents of color would be forced to leave or not have access to parks or other recreational areas.

Variations in political systems had a limited effect on residents' activism and strategies in Dudley, Casc Antic, and Cayo Hueso. In the late 1980s, the Castro regime responded to civic pressure for reform and began to allow Cubans some autonomy in planning and implementing neighborhood projects. Independent and radical initiatives such as the Callejón de Hamel achieved legitimacy on their own with a great deal of international media and artist support. In Barcelona, Franco's dictatorship ended in 1975, and a dynamic civil society—often led by architects who ended up occupying high positions in the municipality in the 1980s—began proposing urban redevelopment projects. However, local residents later protested against municipal and business initiatives. Through the years, they have remained engaged in contesting inequitable public policies in their neighborhood. In Boston, which has a well-established tradition of local democracy, minority and marginalized residents in the 1980s tried to persuade city officials to hear their concerns and respond to abuses in their neighborhoods, but they often acted on their own when political leaders remained passive. The city's verbal commitment to community engagement in historically marginalized neighborhoods such as Dudley actually often benefited private investors and a few city leaders. In short, the three neighborhoods displayed different levels of activism, but residents organized independently in the city and fought to have their demands and engagement recognized by planners and officials.

Residents in Dudley, Casc Antic, and Cayo Hueso were not a monolithic block with uniformly similar values, interests, visions, and tactical preferences. This book has focused on active residents, community leaders, neighborhood workers and their supporters and two decades of their environmental revitalization actions. But not all residents and community workers became involved in urban farms, gardens, farmers' markets, playgrounds, and community center planning. Not everyone in a neighborhood, region, or country is interested in participating in civic action. Even so, activists in each neighborhood tried to promote greater participation from a variety of residents by gathering volunteers for clean-up and maintenance tasks and for protests. Their goal was to build momentum for long-lasting community-based revitalization, and they gathered substantial internal support over time for their projects.

Despite the efforts of community activists to draw residents together, enhance social coexistence, and address internal tensions within ethnic and cultural groups, divergences remained within each neighborhood. In Dudley, conflicts arose around the management of the community gardens, and when ethnic groups did not manage to work out their differences, some of them created a new garden where different vegetables are grown and garden supervision reflects other cultural or gender dynamics. Today, different groups control different community gardens in Dudley, and they mark clear subcultural boundaries: the Dacia/Woodcliff Streets Community Garden tends to be an African American garden, and the Leyland Street Community Garden tends to be a Puerto Rican garden. Some female gardeners from the Leyland Street Community Garden also split off from the main garden and created their own Leyland

Street Community Garden where they can make decisions about garden management.

In Barcelona's Casc Antic neighborhood, disagreements emerged around the rebuilding of the Forat de la Vergonya into a permanent space. Some older and more radical activists refused to accept any intervention from the city (even any investment), but more pragmatic and moderate activists agreed to use municipal funding to construct the Pou de la Figuera in 2006 and 2007. According to them, the local taxes that they pay should go to projects such as the Pou. Other disagreements emerged with the construction of the new multisports arena in 2010. Some associations, such as A. E. Cervantes–Casc Antic (AECCA), supported the new space while others (such as youth organizations) criticized its excessive institutionalization. In Cayo Hueso, some tensions remained throughout the years between Afro-Cuban activists and some white community workers within the TTIB workshop. Supporters of the Callejón de Hamel accused some TTIB workers of displaying racist attitudes toward the cultural and social events organized every week around rumba activities (for example, calling them dangerous behavior), even though both camps worked closely together on autonomous community projects.

Finally, internal divergences between activists emerged during coalition building and while constructing tactical repertoires of actions, especially choosing the attitude to take toward officials and city planners in Barcelona and Havana. In the Casc Antic, squatters and older residents disagreed sharply about whether to distrust the community-engagement practices of the municipality and instead solve problems fully on their own or whether to adopt less radical and more pragmatic postures of negotiation and dialogue. In Cayo Hueso, some community leaders were more inclined to work within the socialist system—with groups such as the Asambela Nacional del Poder Popular (National Assembly of the People's Power) or the Unión Nacional de Escritores y Artistas de Cuba (UNEAC) (National Union of Cuban Artists and Writers)—but others refuse to engage with official Cuban organizations. Such tensions showed that coalitions are malleable, with common interests and values coming together at crucial moments of project development (for example, protecting the Forat as a green space or making structural improvements to the Callejón de Hamel) but with narrower preferences also defended at times by groups and individuals.

Over the years, these internal tensions have required neighborhood leaders and community workers to balance celebrating the neighborhood's ethnic and cultural diversity with strengthening its cohesion.

Many focused on building new leaders, remembering a vibrant past while recognizing recent losses, and encouraging residents to remain unified against new municipal and corporate developments. Events and activities within environmental projects encouraged residents to focus on a single agenda for building new relationships and enhancing resilience against outside pressures.

A Refined Understanding of Urban Environmental Justice: Addressing Physical and Mental Health

The literature on environmental justice has examined the disproportionate environmental burdens that are suffered by residents of low-income and minority communities, who are exposed to more environmental toxins and other health risks than their white and wealthier counterparts are (Bryant and Mohai 1992; Bullard 1990, 2005; Downey and Hawkins 2008; Pellow 2000; Schlosberg 2007; Varga, Kiss, and Ember 2002). To a similar extent, the lands and health of poor and minority populations in the global south have suffered from mercury spills from mines, oil and timber extraction, deforestation and erosion from extensive farming, and hydroelectric dams (Ahmad 1999; Brysk 2000; Carruthers 2008; Evans, Goodman, and Lansbury 2002; Hilson 2002; Martínez Alier 2002). Despite some exceptions that typically focus on the United States (Agyeman, Bullard, and Evans 2003; Gottlieb 2005, 2009; Pellow and Brulle 2005), most environmental justice studies ignore ongoing broad efforts to revitalize neighborhoods and achieve long-term improvements in environmental quality. Residents and their supporters come together for more than protests against isolated threats. Environmental justice publications, such as the *Environmental Justice Journal*, prioritize studies that look at the "adverse and disparate environmental burden impacted marginalized populations and communities" rather than studies that look at recent environmental justice actions taken by those populations.

The efforts, narratives, and strategies of marginalized urban communities that are fighting for environmental justice continue beyond struggles against clearly identifiable contamination sources. This book has looked at a variety of health and environmental improvements from the perspective of the residents and organizations working on the ground. Their vision provides a new lens on improved environmental conditions. The work of activists reveals that some unexpected improvements—such as enclosed community spaces that provide safe opportunities for physical activity and socializing and cafés that sell healthy food and offer a place to build relationships—contribute more to improve environmental value for residents than beautiful, well maintained, but unused parks and playgrounds. Constructing a new sports complex in place of a vacant lot, where residents will gather and practice basketball or soccer, can be envisioned as vulnerable people's idea of getting a green space, as many activists point out.

Public art projects may not seem green at first glance, but they can improve the long-term livability of indoor and outdoor spaces and provide safe and welcoming common spaces. They can be associated with the repair of unsafe neighboring buildings and alleys and with artistic benches and fountains built with recycled materials. Such initiatives cannot be labeled as nonenvironmental because their focus is not on eliminating a toxic factory or improving access to urban forests. Activists interpret the notion of habitat broadly. Their work enhances environmental conditions both inside and outside the house—both green building and sanitation improvements as well as parks and playgrounds. They are remaking a home for residents in the neighborhood.

Community projects in Dudley, Casc Antic, and Cayo Hueso incorporated aspects of urban environmental justice, environmental health enhancement, and broader community development throughout the neighborhood. In that sense, green projects are only the tip of an iceberg that includes equitable and sustainable community development in the form community centers, housing, food, community activities, and economic opportunities. Environmental care, economic development, and social protection are interwoven and need to consider land use, economic development, training opportunities for youth, leadership development, and environmental protection. Activists work with residents for neighborhood sustainability, stability, and resilience and seek solutions that will prevent future degradation and unaffordability.

Neighborhood projects can rebuild a broken community, help residents heal from loss and violence, create safe havens and refuges, and build a diverse urban village with aspects of self-sustainability. In this vision, protection, healing, resilience are the core psychological components of urban environmental justice work. Safety can be conceived of in innovative ways. Safety is about overcoming trauma linked to loss and fear of erasure; recreating rootedness by reviving traditional practices, especially those related to food; creating safe havens and refuges where residents feel comfortable and protected from the stress of the city; addressing negative images about oneself and the neighborhood; deterring environmental violence and dumping; and preventing crime and other unwanted behavior. Through safe havens, residents can find psychological support to rebuild and move forward. In places such as Boston's Body by Brandy gym and community gyms advocated for by Project Right, children exercise in a nurturing space, make positive goals for their future and their families, and learn how to become leaders.

The work of active residents, community leaders, and neighborhood workers in Boston, Barcelona, and Havana shows that urban environmental justice has physical and psychological dimensions and that environmental recovery is achieved by overcoming environmental trauma. Urban sustainability becomes refined by going beyond poverty alleviation and job creation to focus on community rebuilding, place reconstruction, and social coexistence. Safety goes beyond protection against physical, social, and financial damage to include cohesiveness, wholeness, and nurturing. Environmental justice thus creates social sustainability, protects and strengthens the social and human infrastructure of the place, and creates a sense of protection and safety.

An ecosystem health perspective on environmental justice looks at the ways that social and physical environmental factors affect health and that people participate in their own dynamic social and physical ecosystems. Comprehensive community initiatives in the Dudley, Casc Antic, and Cayo Hueso neighborhoods of Boston, Barcelona, and Havana took an ecosystem approach by promoting positive changes in individual, family, and community circumstances and by improving physical, economic, and social conditions (Spiegel, Bonet, Yassi, Molina, Concepcion, and Mast 2001; Yassi, Mas, Bonet, Tate, Fernandez, Spiegel, and Perez 1999). A whole-system perspective is formed at the intersection of basic needs such as food, energy, and shelter in an environmental respectful way. Communities become more resilient and robust as they add the concept of wellness to the equation of environmental revitalization.

In sum, the comparative research conducted in Havana, Barcelona, and Boston shows that the environmental component of environmental justice is fundamental, despite claims that spatial justice is an encompassing category of all forms of justice claims in the city (Soja 2009). In this book, I have strengthened existing bridges between planning and public health scholarship through a qualitative examination of the multiple dimensions of health and environmental health present in environmental revitalization endeavors in traditionally distressed communities. I also have attempted to present a new framework for studying urban and place-based environmental justice action (figure 7.1).

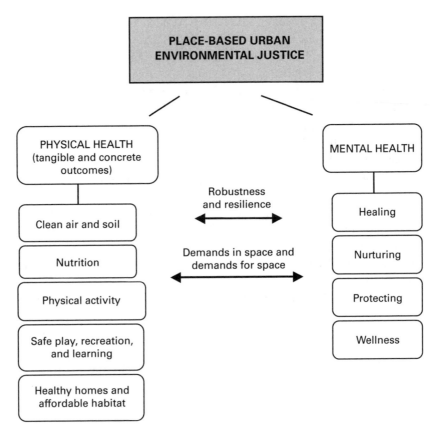

Figure 7.1
A new framework for place-based urban environmental justice

Limits and Future Research

A common criticism that is raised about case study research is that findings cannot be generalized to an entire population of individuals, families, or—in this case—neighborhoods or cities. Case studies are seen as providing only a weak basis for scientific generalization and as having a limited external validity (Yin 2003). Would the questions asked in this book be answered in similar ways if I had studied Bangkok, Moscow, and Paris rather than Barcelona, Boston, and Havana? In fact, case studies are generalizable to theoretical propositions but not to entire populations (Yin 2003). Case studies are not part of a statistical sample, and I am not generalizing my findings here to an entire group of cities or neighborhoods in the global north and south. Rather than enumerate frequencies, the three cases discussed in this book expand and generalize existing theories of environmental justice, the relationship between place attachment and community-engagement scholarship, and urban change and social movement. In that sense, the theory of community mobilization for improved environmental quality in marginalized neighborhoods that led me to choose Dudley, Casc Antic, and Cayo Hueso as case studies is the same theory that can help to identify other cases in which the results are generalized.

At first glance, the three places chosen for the case studies in this book seem to be very different. Havana, in fact, is part of a socialist, centralized, and autocratic regime. But I selected them as part of an inductive approach to research because I was struck by similarities in experiences, discourses and tactics among activists in the three places. I soon realized that a comparative analysis of their mobilization over time would be compelling because residents in different political systems (old democracy, young democracy, and dictatorship) and contexts of urbanization (developed, in process of redevelopment, and developing) took similar types of actions and showed similar commitment to their neighborhood. Contrary to common wisdom and existing studies, my findings show that activists in the three neighborhoods followed similar patterns and dynamics to address marginalization issues.

The common patterns across neighborhoods and cities that were found during the research for this book demonstrate the value of cross-national and longitudinal comparisons. Traditional international development scholarship and urban studies have privileged comparisons across similar socioeconomic development contexts or political systems and have only recently started to compare across development settings (Inam 2005). Conventional scholarship has often neglected to consider how neighborhoods and communities in dissimilar economic, cultural, and political settings mobilize against similar hardships and challenges and what commonalities researchers might find across cases to explain similar positive outcomes or processes. Here I show how activists across a spectrum of development conditions and democratic settings achieved improvements in socioenvironmental conditions, leveraged power with planners and decision makers, and used their visions as a basis for questioning broader political and institutional systems.

This study identified three emblematic neighborhoods that were mobilizing for improved environmental quality. This selection criterion makes the analysis of other historically marginalized neighborhoods in Boston,

Barcelona, and Havana compelling from a control standpoint. It might have been interesting to introduce a control case in each city. A different study could ask why Dudley residents took swift action to revitalize their neighborhood and why residents in nearby Mattapan, another low-income and distressed neighborhood in Boston, did not (until recently). On a different scale, why did residents in Boston, Massachusetts, assemble forces to revitalize marginalized neighborhoods but residents in Portland, Maine, did not focus to the same extent on environmental problems? Are the conditions of abandonment and degradation so extreme in the Casc Antic and Dudley that activists had no other choice than taking action? I argue that conducting this kind of qualitative study with control cases would not be fully relevant because asking individuals why they did not do something or why they focused on one problem rather than another might not necessarily lead to rich and analyzable answers. People have many reasons for not taking action. In this research project, I focused on why people took action and how their actions, meanings, and strategic engagement varied, and I did this by comparing cities across a variety of political systems and urbanization contexts.

It would be interesting to conduct a comparative analysis of the inner-city neighborhoods studied in this project with neighborhoods located in the outskirts of Boston, Barcelona, and Havana. The goal would be to understand how goals (community reconstruction around environmental revitalization) and strategies (of both residents and organizations) are affected by living or working in transitional, suburban, or unstructured geographical settings where community attachment might be less deeply grounded than in historic downtown areas. In Cayo Hueso, Casc Antic, and Dudley, residents are attached to the history, past, and activism of their centrally located neighborhood. Such roots might be weaker or different in suburban communities, informal settlements, or favelas areas. It is worth asking if different marginalized individuals and groups use similar symbols, stories, and imagery to address environmental inequities, contest spatial exclusion, and oppose existing stigmas. Are the initiatives and strategies that are developed by different low-income communities influenced by spatial and historical processes? Such studies would shed new light on the relationships among space, identity, agency, and community organization in a variety of spatial settings.

This book provides a historical and longitudinal perspective on community awareness of and mobilization against environmental degradation and abandonment. Another interesting research study would be to analyze the processes by which residents form emotional connections—how those connections are formed, evolve, last over a lifetime, and inspire some residents and leaders to work for their neighborhoods. To date, the research on community connections has tended to take place in the environmental psychology field, which has focused on individual experiences and meanings rather than collective ones. However, the planning and environmental justice literature tends not to examine personal experiences of place and attachment and instead limits analysis to neighborhood-level processes and external forces that affect neighborhoods. Further bridging these fields could lead to fruitful new knowledge and help policy makers plan more just and meaningful environmental projects.

Implications for Policy and Planning

Planning Dilemmas
When faced with the kinds of civic demands that were formulated by activists in Barcelona, Boston, and Havana, policy makers confront two difficult dilemmas. First, how to balance demands for safe havens and minimize risks of self-segregation. As community members attempt to protect residents and control their land and its borders, they also build a self-sustained and contained community with a limited membership. To some extent, they reprivatize public spaces for their own uses and make their territory exclusive, and this happens with the support of local planners and politicians. This is especially true in Dudley and Casc Antic. Residents who want to live in a homogeneous and familiar community seek to protect themselves from alienation and threats to the neighborhood and also prefer to live among residents from similar ethnic and cultural backgrounds. Racially homogeneous spaces help them strengthen a shared racial identity, can provide a soothing oasis away from the pressures of interethnic relations, and can help residents bolster their ability to deal with negative racial relationships (May 2001).

Yet, such homogeneity also can threaten the ideal of diversity and the "good city" by creating pockets of isolation throughout the city. The question of how planners can best balance needs for protection and diversity remains open. Self-segregation tendencies might be able to be addressed by considering the types of threats that residents fear most—encroachment, gentrification, displacement, and exclusion—and possible remedies. Policy makers and planners should focus on strengthening housing affordability policies. In the United States, local governments and federal agencies (the U.S. Department of Housing and Urban Development, in particular) will need to ensure that good-quality, affordable housing rentals and sales are provided to residents from all ethnic backgrounds and that local policies are in place to secure such housing. In Spain, cities will have to increase the number and quality of official housing protection units and strengthen the role played by housing cooperatives in neighborhoods in transition to ensure that affordable units remain available to low-income residents. In both countries, housing cooperatives are a model for affordable and sustainable housing that should be explored to a greater extent. Public housing should be attractive and well made, which will help residents feel less stigmatized in their territory. Finally, in Cuba, the government should ensure that the new *permuta* system of housing exchanges and sales does not create territorial inequalities in Havana by which wealthier residents move into the center city and displace poorer residents. International funding should also be sought so that neighborhoods such as Centro Habana benefit from the same level of renovation as neighborhoods such as Habana Vieja.

In all contexts, cities should support community-based organizations that offer environmental, social, and cultural services and projects to residents. Programs for youth residents provide mentoring and after-school support before young people return home for the evening. After-school programs—such as the ones developed by Fundació Comtal and A. E. Cervantes - Casc Antic (AECCA) in Barcelona, Quiero a Mi Barrio in Havana, and Alternatives for Community and Environment (ACE) in Boston—are crucial to ensuring community cohesion, youth development, and a strong network of caring adults who can offer community learning and mentoring opportunities. Finally, municipalities should provide administrative support and progressive fiscal policies that help minority and low-income residents create businesses. This would help to build a local economy that provides new jobs in the community and allows residents to access affordable and quality goods and services directly in their neighborhood. These businesses include small local supermarkets, welcoming social venues related to food, and home and garden improvement stores.

The second dilemma that planners face is related to questions of urban sustainability: how can planners help make cities more livable and environmentally strong and also satisfy community demands for memory and rootedness? In Dudley, Casc Antic, and Cayo Hueso, activists and residents struggled to define the role that memory should play in urban sustainability. Dense urban neighborhoods often do not have enough empty space for new parks, playgrounds, and urban farms, especially when housing is in demand. Producing new environmental goods often involves demolishing existing buildings and spaces that minority residents feel strongly connected to and whose erasure triggers feelings of loss. Decisions over land uses in historically marginalized neighborhoods are complex, and the goal is to achieve solutions that improve living conditions and also consider the place of memory and identity.

A possible answer comes from activists' demands for self-management and spontaneous participation in neighborhood initiatives. If people feel that they themselves can lead environmental projects and achieve this balance between environmental sustainability and memory, they might feel less uprooted from the space. The United States has environmental justice laws and grants, so funders could incorporate criteria of cohesiveness, nurturing, and protection into their decisions to support specific projects. Funders in other countries should also ensure that dimensions such as community identity and place attachment are not relegated to the background during the planning process. In addition, planners could direct more funding to revitalization projects that address the psychological dimensions of environmental health, especially as they relate to trauma, recovery, and safety.

Avoiding Silo Thinking about Environmental Justice

In Barcelona, Boston, and Havana, activists engaged in a variety of far-reaching environmental revitalization projects and showed that environmental justice and community development are closely linked. They developed urban farms, community markets, parks, small green spaces, playgrounds, recreational fields, multipurpose community centers, healthy and green housing complexes, job opportunities, and welcoming venues for healthy food and community activities. In traditionally forgotten urban communities, planners and decision makers need to go beyond eradicating pollution sources and waste sites and promoting the redevelopment of empty lots by private investors. These communities present value and often have strong social and spatial capital, and redevelopment projects should not fit the white idealized notion of place and neighborhood. Traditional "brown" concerns of the EJ movements pointed at policymakers and dominant institutions for treating marginalized groups as if they do not deserve to live in healthy neighborhoods, and allowed the siting of waste sites where they lived. Green gentrification would be the flipside of that process, through which the urban poor and people of color are only allowed to live in less healthy and livable neighborhoods.

Furthermore, to date, planning practices tend to remain compartmentalized and to lack a holistic vision for environmental justice and community revitalization. Based on the experiences of residents in Cayo Hueso, Dudley, and Casc Antic, environmental-revitalization projects in cities need to coordinate urbanism, territorial planning, environment, economic development, public health, and youth and recreation. Funding for community groups and organizations should be better coordinated by program managers who are engaged with these municipal departments and respected in urban neighborhoods. Current planning practices often overlook holistic visions of environmental justice, place remaking, and healing and instead focus on concrete and measurable outcomes, even though marginalized communities reap great emotional and psychological benefits from socioenvironmental projects.

Finally, residents and their supporters are attempting to invigorate existing democratic practices at the local level. Top-down planning and policies have distorted deliberative democracy discourse, disengaged planners from the communities, and been rooted in the growth machine. The experiences of activists in the three neighborhoods in Havana, Boston, and Barcelona reveal that informal, spontaneous and community-led processes can result in successful and more innovative planning. Democracy does not necessarily mean deliberative democracy or protests and social movements. Sometimes local democracy can experience cycles of autonomous and at times anarchical planning, protests, negotiating, and neighborhood actions. Such iterative processes help in situations where power differentials and challenging intercultural and socioeconomic relations are central elements for planning in distressed urban neighborhoods.

Final Thoughts

Over the past two decades, the Dudley, Casc Antic, and Cayo Hueso neighborhoods of Boston, Barcelona, and Havana have achieved many environmental and health successes. Residents and their supporters have collaborated to provide welcoming, secure, green, and equitable neighborhoods where children, young people, and older residents can live, play, and eat healthy food. They also are united in shaping the goals and skills of the next generation of residents so that they help sustain the current successes over time. When younger people leave, older leaders pass away, and municipal or national dynamics (in Cuba in particular) change, residents need to have options for action. They need to understand how to stave off gentrification and displacement of residents. In ten years, researchers will need to revisit places such as the Leyland Street Community Garden, The Food Project farms, the Callejón de Hamel, the Quiero a Mi

Barrio gym, and the Pou de la Figuera to assess whether the vision of local activists has been preserved and even strengthened over the years.

In this book, I have proposed a new framework for studying place-based urban environmental justice action based on the initiatives and projects that active residents, local leaders, neighborhood workers, and their supporters created in Dudley, Casc Antic, and Cayo Hueso. These projects included urban farms, community gardens, farmers' markets, parks, playgrounds, small green spaces, sports grounds, community centers, healthy homes, and improved waste management that were built thanks to bottom-to-bottom networks. For over two decades, activists in Boston, Barcelona, and Havana have worked to improve the environmental quality and livability of their neighborhoods through projects that have transformed socioenvironmental conditions, rebuilt broken communities, and remade place for residents.

I have argued that a theory of environmental justice for urban neighborhoods needs to include specific outcomes and processes. Outcomes take the form of improvements in physical health, such as clean air and soil; healthy and affordable nutrition; safe play, recreation, and education; physical activity and sports; and healthy and affordable homes and habitat. Outcomes also include mental health support: in degraded, abandoned, and traumatized neighborhoods, families, youth, and children need nurturing, healing, protection, and wellness. Over time, improved physical and mental health outcomes create greater community robustness and resilience. In addition, several processes are needed for communities to achieve urban environmental justice, including addressing stigmas about low-income and minority residents, controlling neighborhood land and territory, building and protecting borders, promoting forms of spontaneous planning and transgressive participation, and protecting spatial capital and place identity. Under these conditions, we can create a healthy environment where all people live, work, play, and learn.

Appendix A
Semistructured Interviews: Barcelona

2008, 2009, and 2012

Albert, Grup Ecologista del Nucli Antic de Barcelona

Pere Cabreras, architect and formerly at in charge of an Area de Rehabilitación Integral

Carola, Equipament del Pou de la Figuera

Emili Cota, Plataforma de Entidades de la Ribera

Pep Dalmau, Grup Ecologista del Nucli Antic de Barcelona

Paco del Cuerpo, Colectivo del Forat

Emanuela, Arquitectos Sin Fronteras

Esperanza, Associació Los Rios en Catalunya

Eva, Arquitectos Sin Fronteras

Jordi Fabregas, Convent de San Agusti

Chema Falconnetti, film director

Jeremie Fosse, Ecounión

Rosa Garriga, municipality of Barcelona, District Ciutat Vella

Gemma, Equipament del Pou de la Figuera

Francesc Giro, Acción Natura

Carme Gual, Foment de Ciutat Vella

Enrique Ibanez, Associació de Veïns en Defensa de la Barcelona Vella

Isabel, Hortet del Forat

Joan, Herboristeria and Associació de Veïns en Defensa de la Barcelona Vella

Jordi Jovet, Fundació Comtal

Khadeja, Morroccan mother and volunteer

Laia, Recursos d'Animació Intercultural

Manolo, Hortet del Forat

Marc, Fundació Casc Antic

Maria, Hortet del Forat

Marta, Ecoserveis

Isabel Martinez, architect and member of Icaria

Oscar Martinez, Cooperativa Porfont

Maria Maso, Associació de Veïns del Casc Antic

Dani Mateos, Associació Portal Nou

Melanie, Eica Espai d'Inclusio y Formacio Casc Antic

Merche, A. E. Cervantes–Casc Antic

Jordi Milaro, Xarxa de Consum Solidari

Pep Miro, Associació de Veïns del Casc Antic

Mohammed, volunteer at Hortet del Forat

Mónica, squatter

Montse, A. E. Cervantes–Casc Antic

Eduardo Moreno, lawyer

Elisenda Ortega, Municipality of Barcelona

Silvana Ospina, Incita

Paco, Hortet del Forat

Hubertus Poppinhaus, social architect and member of Associació de Veïns en Defensa de la Barcelona Vella

Rafa, formerly at Adsis

Miguel Rene, Municipality of Barcelona

Jorge Sanchez, Federació d'Associació de Veïns i Veïnes de Barcelona and Associació de Vecinos Casc Antic

Marc Aureli Santos, municipality of Barcelona

Victor (name changed), Associació de Comerciantes Calle Carders

Virginia, Companya Moki Moki

Appendix B
Semistructured Interviews: Boston

2009 and 2012

Alexandria King, The Food Project
Andrea Taafe, director of the Shirley-Eustis House
Melida Arredondo, Upham's Corner Health Center
Andrew Bailey, Riley Foundation
José Barros, Dudley Street Neighborhood Initiative
Peter Bowne, Earthworks
Bing Broderick, Haley House Bakery Café
Marcia Butman, Discover Roxbury
Pansy Carlton, Leyland Street Community Garden Extension
Lisa Chapnick, formerly at the Boston Department of Human Services
Dawn Chavez, Boston Youth Environmental Network and Dearborn School Garden
Corinna (name changed), Savin and Maywood Streets Community Garden
Brandy Cruthird, Body by Brandy
Jeanne Dubois, Dorchester Bay Development Corporation
Andria Post Ergun, Boston Department of Neighborhood Development
Alma Finneran, Virginia and Monadnock Streets Community Garden
Drew Forster, Kroc Foundation
Evelyn Friedman, Boston Department of Neighborhood Development and formerly at Nuestra Communidad Development Cooperation in Dudley
Aldo Ghirin, Boston Parks and Recreation Department
Alice Gomes, community organizer

Honorio, Cape Verdean leader

Bo Hoppin, Boston Youth Environmental Network and Dearborn School Garden

Betsy Johnson, South End Land Trust

Charlotte Kahn, former member of the Boston Urban Gardeners

Mel King, political activist

Mike Kozu, Project Right

Carmen LaTorre, Jardín de la Amistad

Tunney Lee, Massachusetts Institute of Technology

Jess Liborio, The Food Project

Penn Loh, formerly Alternatives for Community and Environment

May Louie, Dudley Street Neighborhood Initiative

Sister Margaret, Project Hope

Anne McHugh, Boston Public Health Commission

Kathe McKenna, Haley House Bakery Café

Nelson Merced, formerly at the Alianza Hispana

Tubal Padilla, former Puerto Rican leader

Marlo Pedroso, Boston Natural Areas Network and Boston Gardeners' Council

Susan Redlych, former member of the Massachusetts Corporate Wetlands Restoration Partnership and volunteer at The Food Project

Eric Seaborn, Massachusetts Department of Conservation and Recreation, Urban and Community Forestry

Trish Settles, former member of the Dudley Street Neighborhood Initiative

Julie Stone, Boston Schoolyard Initiative

Paul Sutton, Boston Parks and Recreation Department

Vidya Tikku, Boston Natural Areas Network

Chuck Turner, former Boston city councilor

Father Walter Waldron, St. Patrick's Church

John Walkey, Urban Research Institute

Greg Watson, formerly at the Dudley Street Neighborhood Initiative (1995 to 1999) and commissioner of the Massachusetts Department of Agricultural Resources

Travis Watson, Dudley Street Neighborhood Initiative

JoAnn Whitehead, Boston Natural Areas Network

Appendix C
Semistructured Interviews: Havana

2009

Adys, writer

Gisela Arandia, Taller La California

Rafael Betancourt, Instituto Canadiense de Urbanismo

Martha Chan, Instituto Nacional de Higiene, Epidemiologia y Microbiologíab

Clara (named changed), State Market, Cayo Hueso

Mario Coyula, Grupo para el Desarrollo Integral de la Capital (GDIC)

Miguel Coyula, GDIC

Cristián, instructor of tae kwan do at Beisbolito

Caridad Cruz, Fundación Antonio Nuñez Jimenez de la Naturaleza y el Hombre

Daniel (name changed), GDIC

David (name changed), Centro Felix Varela

Deyni, Proyecto Moros y Cristianos

Joel Diaz, Talleres de Transformación Integral del Barrio, Cayo Hueso

Eduardo, volunteer at the urban farm

Elias, Callejón del Hamel, Cayo Hueso

Maria del Carmen Espinoza, Talleres de Transformación Integral del Barrio, Cayo Hueso

Evaristo, Office for Urban Agriculture, Municipality of Havana

Armando Fernandez, Fundación Antonio Nuñez Jiménez de la Naturaleza and el Hombre

Jesus Figueredo, Martin Luther King Center

Franco (name changed), Oxfam

Froilan, writer

Arsenio Garcia, United Nations Children's Fund

Gladys, permaculture practitioner

Laritza Gonzales, Centro de Intercambio y Referencia Iniciativa Comunitaria

Graziella, Federación de Mujeres Cubanas

Rafael Hernandez, Instituto Cubano del Arte y la Industria Cinematográficos

Noemi Reyes Herrera, Taller de Pogolotti

Iñaki, Save the Children

Jaime (name changed), Quiero a Mi Barrio

Felix Janes, Martin Luther King Center

Mujeres Jardineras, Barrio de San Isidro

Jesus "Lara," artist

Cecilia Linares, Centro Juan Marinello

Luciano, permaculture practitioner

José Luis, urban farm, Cayo Hueso

Laura (name changed), Plan de Rehabilitación Urbana del Municipio de Centro Habana

Roxana Mar, physician, Cayo Hueso

Mario, community leader, Cayo Hueso

Modesta, Espada 411, Cayo Hueso

Olga, Espada 411, Cayo Hueso

Reina (name changed), Grupo para el Desarollo Integral de la Capital (GDIC)

Pablo (name changed), Casa del Nino y Nina, Cayo Hueso

Pamela (name changed), Ayuda Popular Noruega

Ester Perez, Martin Luther King Center

Estela Rivas, historian of Centro Habana

Rosa, Casa del Nino y Nina, Cayo Hueso

Salvador, Callejón de Hamel, Cayo Hueso

Jerry Spiegel, University of British Columbia

Joel Suarez, Martin Luther King Center

Annalee Yassi, University of British Columbia

Appendix D
Techniques for Data Analysis

The core of my data analysis was based on interviews that I conducted in the Dudley, Casc Antic, and Caya Hueso neighborhoods of Boston, Barcelona, and Havana. I met with activists (active residents, leaders, community workers), their supporters (foundations, nongovernmental organizations, informal groups), public officials, and municipal staff members. Together, these interviews, my observations, participant observations, and secondary documents provided rich empirical data for the analysis of environmental revitalization in these three neighborhoods.

During each interview, observation, and participant observation, I took extensive nonselective notes reflecting the respondents' accounts, and after each day in the field, I used those notes to develop analytical short memos. Those memos summarized the context of the interview and my first impressions from the data, organized the notes into themes, and helped me process some of the data. As I was conducting my fieldwork, I regularly wrote longer analytical memos to reflect some patterns encountered in the data as they related to my core research questions. In addition, I built initial models showing relationships between themes and showing relationships among the interviewees' visions, objectives, and strategies around environmental justice projects. Approximately three quarters of the interviews were fully transcribed, and for the remaining one quarter, I transcribed my extensive notes. I fully transcribed all the interviews that seemed essential to my study and partially transcribed interviews that provided a general context for the research or that were very brief.

At the end of my fieldwork, I engaged in a more thorough and systematic data analysis. I used grounded theory techniques to analyze press articles, reports, documents produced by organizations, interviews with planners and decision makers, and interviews with key local leaders, neighborhood residents, and community workers (whose active engagement had a great effect on the neighborhood's access to environmental and health projects). These techniques included coding text systematically to avoid the application of predefined concepts on the data (Glaser and Strauss 1967; Charmaz 2006). This work involved completing two layers of line-by-line and paragraph coding—fragmenting, sorting, and separating the entire textual data systematically into categories. In a second stage of focused and axial coding, I used the most significant and frequent codes to synthesize, integrate, and organize my data into theoretical concepts. I related categories to subcategories and reassembled the data I had fractured during line-by-line and paragraph coding to bring more coherence to the emerging analysis. I did two rounds of coding for each interview to ensure that I was thoroughly recording concepts and integrated excerpts of interviews that reflected codes I created at a later stage. After completing this coding, I wrote memos and theorized about the relationships between concepts, attempting to link the theoretical patterns I unraveled around a central category that knitted everything together (Glaser and Strauss 1967). I also compared the interviews and looked for similarities and differences between respondents and between neighborhoods. These memos helped me generate models that showed the mechanics and relationships in my findings and allowed me to visualize emergent theories (for an example of this coding and analysis work, see appendix E).

During data coding, I developed codes around notions of marginality, neighborhood decay, domination, justice, participation, contestation, identity, attachment, space, coalitions, and networks, which reflected my initial theoretical assumptions and preliminary findings. However, my coding techniques also allowed for the emergence of unexpected concepts, including self-management of public space, sense of responsibility toward neighborhood, protection of patrimony, defense of existing social fabric, nostalgia, emotional fulfillment, war zone, safe haven, nurturing, bricolage, and quick wins. This grounded theory helped me build stories of local leaders, understand their individual and collective identities, and unravel their engagement in the neighborhood and their vision for its revitalization. It was meaningful for understanding the dual role that was played by place in local activism. Finally, it allowed me to look in depth at the narratives and strategies that leaders developed over time to organize residents and gather support and also the type of support (or lack of support) and engagement that was provided by outside organizations, planners, municipal staff, and policy makers.

In parallel, I used process-tracing techniques for the analysis of interviews (with local leaders, active residents, nonprofit organizations, and policy makers and planners), original documents, and notes from observations and participant observations. Process tracing involves evaluating evidence about the causal mechanisms that link causal variables to outcome variable, searching for the ways through which causal variables are related to the outcome variable (Brady and Collier 2004; George and Bennett 2005; Gerring 2007). It allows for the understanding of complex causal relationships such as those exemplified by multiple causality, feedback loops, and complex interaction effects. It directed me to trace processes in a specific and theoretically informed manner as I mapped the process, explored the extent to which it coincided with prior, theoretically derived expectations about the workings of the mechanisms, and focused on specific decision making dynamics. Eventually, I was able to formulate new minitheories on the causal mechanisms that connected correlated phenomena in my research (Checkel 2005). Particularly important in my analysis was the explanation of decision processes by which initial conditions and situations of marginalization and neighborhood degradation became translated into outcomes and were mediated by different stimuli (inside or outside the person's environment or institutional arrangement) (George and McKeown 1985).

To ensure that my process tracing was exhaustive, as I coded the data, I developed codes and processes that captured and pieced together the elements needed to answer my research questions. Through process tracing, I analyzed how community engagement is built on residents' and organizations' attachment to a neighborhood, their collective identity, broader urban dynamics, and their sense of marginalization and power in the city. I also focused on the development and refinement of strategies and tactics over time and on the factors and external actors that affected (positively or negatively) different choices. This analysis explained how marginalization and fragmentation of environmental spaces, people's sense of place and understanding of their neighborhood, and the local social and political contexts affected the engagement, demands, and strategies of marginalized communities.

Finally, I used a combination of two techniques, analytic and historical narratives. Analytic narratives seek to account for "outcomes by identifying and exploring the mechanisms that generate them" (Bates 1998, 12) based on the "actors' preferences, their perceptions, their evaluation of alternatives, the information they possess, the expectations they form, the strategies they adopt, and the constraints that limit their actions" (Bates 1998, 11). Historical narratives support this approach by recognizing that feedback loops are important in decision making and that historical changes and events reshape the preferences of actors and make previously available repertoires no longer available (Büthe 2002). Time itself becomes an element that is part of the causal explanation. Historical narratives are useful for contextualizing the different steps of the process rather than leaving them fragmented into analytical stages (Checkel 2005).

This combined historic and analytic narrative technique was particularly relevant for understanding how urban development projects, local histories of marginalization, and political opportunities motivated activists in their struggles for environmental goods and livability. This technique also allowed me to assess the importance of institutional changes in the local groups that facilitated residents' work and achieved success in improved environmental conditions. During the coding phase of my data analysis, I built some categories that included these elements and later assembled them into dynamic processes, histories, and explanations.

Appendix E
Constructing Grounded Theory

List of NVivo Free Nodes after Line-by-Line and Paragraph Coding for Understanding the Strategies and Tactics Developed by Neighborhood Activists

This list illustrates the list of free nodes developed in NVivo software through a line-by-line and paragraph coding of interview documents, observations and participant observation notes, and secondary data documents. "Sources" refers to the number of sources (documents) in which a specific free node (such as "Advocacy and lobbying") appears, and "References" indicates the number of total references relevant for this node. A source can have multiple references. Nodes with a larger number of sources and references are generally the most salient ones. This stage of analysis helped me select, sort, and separate data into broader categories. I built a total of thirty free nodes for my analysis of strategies and tactics.

Examples of Quotes within the Free Node "Bricolage and Collage" for Casc Antic, Dudley, and Cayo Hueso

The quotes that are listed below are an excerpt of some of the raw data that I analyzed and incorporated into the free node called "Bricolage and collage." I coded the raw data from interviews, observations and participation observations, and secondary documents through line-by-line and paragraph coding. "Reference" and the number next to it indicate the number of excerpts from a particular source (such as an interview) used for a specific node, such as "Bricolage and collage." "Coverage" and the percentage next to it indicate the percentage of a source (such as an interview) that a particular excerpt or quote covers. The value of the coverage and reference illustrate the salience and importance of an excerpt or quote within a node.

Name	▲	Sources	References
Advocacy and lobbying		23	51
Attachment to neighborhood, its history, its social movements		54	198
Bricolage and Collage		33	102
Celebrations and partes		13	35
Claim discussion and planning during socio-cultural events		21	31
Competition and conflicts between civil society groups and organization		6	16
Confluence of disaggregated interests together from various groups		3	4
Contestation and protest to construction to dialogue as a cycle		23	103
Cultural and art and spiritual dimensions of environmental projects		23	83
Denunciations, complaints, occupations, and provocations		22	51
Environmental education as form of empowerment		24	58
Fundraising, funding, and grants		23	78
Leader (importance of) and role of inspiration and guide		13	45
Leadership and social capital development		53	198
Negotiation and dialogue (emphasis on) and space of discussion and debate		23	55
Organizational strengthening		7	16
Organizing at the community level		31	97
Political opportunities and positive political context and broader public policies		7	28
Politicians need to support neighborhoods for their own benefits		8	11
Pride over improvements and comfort in neighborhood and need to cultivate pride		31	67
Public private partnership and city relinquishing its power and control		9	21
Quick wins as ways to bring neighborhood together		13	27
Reputation as a tool		6	23
Resourcefulness and non-material resources		27	71
Sub-community networks and loose and flexible connections and partnerships		55	227
Technical studies reports and evaluations and documentaries		4	12
Unexpected and broad coalitions		54	189
Use of Place Attachment and Community Identity and Memory and Place characteri		17	59
Volunteers (use of) and their work and dedication		15	23
Youth and Children (organizing around them, using their spontaneity)		15	39

Title row: **Strategies and Tactics Free Nodes**

Figure E.1

Example of NVivo free nodes for data analysis

<Documents\Barcelona\07.16.2009 Virginia Compania Moki Moki> - § 1 reference coded [9.09% Coverage]

Reference 1–9.09% Coverage

Luchamos con nuestras herramientos. Hacemos politica a nuestra manera. Conocmos a RAI con el teatro. Cada uno ponemos nuestro granito de arena.

<Documents\Barcelona\6.18.2009Laia RAI transcribed> - § 1 reference coded [2.24% Coverage]

Reference 1–2.24% Coverage

Hubo un espacio un poco más limpio y más que se había hecho de a poquito que se habían visto crecer las plantitas los bancos, yo que sé, el campo de fútbol.

<Documents\Barcelona\7.14. Hubertus Poppinhaus transcribed> - § 4 references coded [1.65% Coverage]

Reference 1–0.30% Coverage

Hicimos plantadas, plantada general, sábado 16 de noviembre y venía la gente con macetas y plantaban árboles, arbustos.

Reference 3–0.49% Coverage

Entonces y de forma anárquica, había okupas que llenaron de macetas públicas de otros sitios, arrancados y los ponían en el Forat, cada uno como podía... Pero al final lo impresionante es que se logró un espacio verde, es que deberías haberlo visto. Tengo fotos antes de que lo han derribado, tengo las fotos. Y al final estaba así.

<Documents\Barcelona\7.19.2009 Merche y Montse transcribed> - § 7 references coded [7.04% Coverage]

Reference 1–0.82% Coverage

Sí que es verdad que el mogollón lo llevamos entre 3 o 4 madres que somos las que estamos más, porque también a lo mejor, por ejemplo yo me encargo más de todo lo de las subvenciones, como yo me entero y con otra madre que también se llama Montse, nos quedamos a las 5.00 de la noche, a la hora que sea para montarlas para poder subsistir.

<Documents\Barcelona\news articles forat de la vergonya> - § 2 references coded [0.71% Coverage]

Reference 2–0.53% Coverage

Los vecinos convocaron una cadena humana el pasado 11 de abril que impidio el avance de las maquinas que pretendian llevar a cabo el derribo. Todas las asociaciones se oponen al plan del Ayuntamiento e insisten en que nunca han recibido respuesta a sus propuestas alternativas

<Documents\Boston\10.14.2010 Interview with TS Ex DSNI> - § 4 references coded [7.27% Coverage]

Reference 2–2.12% Coverage

Dennis Street Park kind of organized itself. We were just kind of a facilitator after. We were able to bring the resources there that people were kind of demanding anyway. It was just a matter of listening and then being able to kind of um, listening kind of gleaming to what the issues are and at some point if you can bring some resources to bear often they are there to help you kind of point those resources in the right direction. Whether the resources are just stack time to another community, neighborhood clean up or 'look hey we've got money to put into benches' and Dennis

Street park was one of these parks where 'well hey, we've got a resource how do we do it? Let's put benches here and here. Oh, we've got another resource.' And maybe like the resources are tiny, very piece meal for a long time but at some point after doing a piece meal project for a long time it develops enough of kind of a community momentum around it that it really drives itself.

<Documents\Boston\12.10.2010 Jeanne Dubois Dorchester Bay> - § 3 references coded [2.20% Coverage]

Reference 1–0.53% Coverage

So we did a lot of clean ups here and here. We did a lot of tree plantings here and here. We've identified other green spaces. And then we had a, we built a playground, the whole neighborhood built a playground.

<Documents\Boston\9.29.2009 Nelson Merced La Alianza Hispana transcribed> - § 3 references coded [2.83% Coverage]

Reference 1–1.45% Coverage

And that was a strategy you know, similar to the strategy people have in some cities where they board up the houses because they can't do anything then they paint scenery on the houses, on the boarded up plywood. This was the same thing you know, you have all this vacant land so we might as well throw grass seeds and flower seeds and stuff like that and just create a meadow. And so that was the theory at the time and people said that's a good idea. Let's try to clean it up as much as we can and we will do annual cleanups and, the whole dump on us, don't dump on us campaign, really sort of the alternative is once you stop the dumping then you can plant grass and gardens and stuff.

<Documents\Havana\10.28.2010 Rosa, Casa del Niño y de la Niña transcribed> - § 3 references coded [2.23% Coverage]

Reference 1–0.18% Coverage

400 kids wrote a letter in the government gave us the space. We received a donation from UNICEF and parents also helped to repair it. Kids planted seeds.

<Documents\Havana\10.28.2010 Former GDIC Staff 1, 3.13.2010 transcribed> - § 2 references coded [1.5% Coverage]

Reference 1–0.11% Coverage

The crisis changed people's way of thinking, which was that the state was going to resolve everything. So many people evolved toward individualism and looked on their own for the solution to their problems.

Reference 2–0.10% Coverage

Lower middle class people invent things themselves. Today half of the dwellings in Havana are slums. People work in state enterprises only to steal materials. And they work on weekends for private people.

<Documents\Havana\11.02.2010 Coordinator, Quiero a mi Barrio transcribed> - § 3 references coded [2.53% Coverage]

Reference 1–0.14% Coverage

One day, I was in front of my house and noticed that there was a non-utilized space. How was it possible that I did not notice it before? So I started "hacer gestion." Inside, it was worse than a Dracula movie. You could not open the door. There were insects, fecal water, sewage, flooding, and no light. But this did not scare me. I started working there with community students. We did cleaning work. We have not received any urbanistic help. Everyone gave pieces. One brought cement, other something else, and we did it little by little. Now, Save the Children is going to repair the toilets for us. We also have received help from Puerto Rico—they were karate practitioners from Puerto Rico. There is a gym, a workshop for plastic arts. All the gym and the machines in the gym were built by us and the kids.

Focused and Axial Coding: Reassembling Free Nodes into Larger Theoretical Concepts and Patterns

In a second stage of data analysis, I used the most significant and frequent free NVivo nodes to sort, synthesize, integrate, and organize my data into theoretical concepts to refine its meaning and start developing larger theoretical concepts (such as "Inventiveness and creativity"). I related categories to subcategories and reassembled the data I had fractured into free nodes during line-by-line and paragraph coding in order to create more coherence in the emerging analysis. Table E.1 presents a short example of this coding.

Constructing Relationships and Models

The model in figure E.2 summarizes the common tactical repertoire that was used by activists in Dudley, Casc Antic, and Cayo Hueso to achieve community reconstruction. I built it after completing different layers of coding work, pattern construction, relationship building between findings, and process tracing, which I used to unravel the connections between strategies and their development through time and stages.

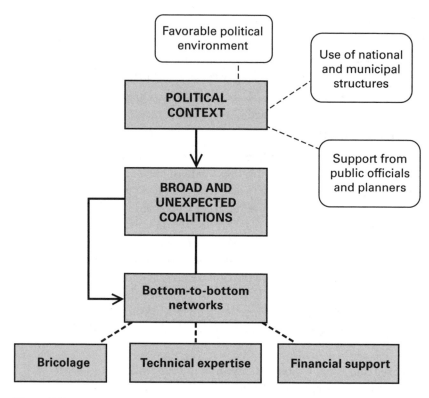

Figure E.2
Example of a model derived from data analysis

Table E.1

Example of a summary table for focused and axial coding

Focused and axial coding categories	Subcategories (based on free NVivo nodes)
Favorable political context	National or local windows of opportunities
	Use of broader political traditions and structures
	Supporters and allies in state or municipality
	Politicians who support neighborhoods for their own benefit
	Clever engagement with officials and planners
Coalitions and networks	Versatile, mixed, and often unexpected coalitions
	Confluence of disaggregated interests from various groups
	Bottom-to-bottom networks with loose and flexible connections and partnership
Activism in practice	Contestation and protests
	Advocacy and lobbying
	Denunciations, complaints, occupations, and provocations
	Bricolage
	Creative collage of materials and resources
	Quick wins as ways to bring together the neighborhood
	Resourcefulness and nonmaterial resources
	Leader's importance and role of inspiration, guide, and guarantee
	Reputation as tool
	Volunteers and their work and dedication
	Youth and children (organizing around them, using their spontaneity)
	Technical studies and expertise
	Grants, funding, and fellowships

Notes

2 Environmental Justice, Urban Development, and Place Identity

1. See, for instance, the growing focus of various journals—such as *Environmental Health*, the *American Journal of Health Promotion* (*AJHP*), the *American Journal of Preventive Medicine* (*AJPM*), and the *Journal of the American Planning Association* (*JAPA*)—on the relationship between elements of the urban environment and health issues (such as obesity, cardiovascular disease, diabetes, lack of physical activity, and asthma). In 2003, *AJHP* released a special issue on "Health-Promoting Community Design" (vol. 18, no. 1) that looked at the built environment and public health. In 2008, *AJPM* published a supplement on neighborhood design and active living (vol. 35, no. 6). In 2006, *JAPA* addressed the role that planning plays in building healthy cities (vol. 72, no. 1).

2. For examples, see the Web sites of some environmental justice organizations in the United States: Alternatives for Community & Environment (ACE-EJ) (http://www.ace-ej.org), Communities for a Better Environment (http://www.cbecal.org), Power U (http://www.poweru.org), Environmental Health Coalition (http://www.environmentalhealth.org), West Harlem Environmental Action (http://www.weact.org), Detroiters Working for Environmental Justice (http://www.dwej.org/about-us), and the New York City Environmental Justice Alliance (http://www.NYCEJA.org). At the global level, see the Web sites of important environmental justice networks and organizations: the Southwest Network for Environmental & Economic Justice (http://www.sneej.org), the Asia Pacific Environmental Network (http://www.apen4ej.org), the Indigenous Environmental Network (http://www.ienearth.org), Amazon Watch (http://amazonwatch.org), the Latin American Observatory of Environmental Conflict (www.conflictosmineros.net), and the Environmental Justice Networking Forum (http://www.bcb.uwc.ac.za/inforeep/EJNF.htm).

3. For more information, see http://www.righttothecity.org/our-history.html.

4. The vision for the recently created journal *Environmental Justice* confirms these trends and priorities: "The *Journal* explores the adverse and disparate envi-

ronmental burdens impacting marginalized populations and communities all over the world. The *Journal* addresses (a) studies that demonstrate the adverse health effects on populations who are most subject to health and environmental hazards, (b) the protection of socially, politically, and economically marginalized communities from environmental health impacts and inequitable environmental burden, (c) the prevention and resolution of harmful policies, projects, and developments and issues of compliance and enforcement, activism, and corrective actions." For more details, see http://www.liebertpub.com/overview/environmental-justice/259.

5. Existing studies (mostly in environmental psychology and not based in urban distressed areas) argue that people's attachment to nature—and not people's civic attachment to a place—is closely associated with pro-environmental behavior (Scannell and Gifford 2010).

3   Stories of Neighborhood Abandonment, Degradation, and Transformation

1. The Dudley Street Neighborhood Initiative has been well documented elsewhere (Layzer 2006; Medoff and Sklar 1994; Putnam, Feldstein, and Cohen 2004; Shutkin 2000). Here I provide a comprehensive story of neighborhood environmental revitalization efforts.

2. For more information, see http://www.bostonnatural.org/communitygardens .htm. Over the years, the Boston Natural Areas Network (BNAN) has helped over ten thousand gardeners plan gardens, acquire property, and design and manage initial capital improvements. It oversees 150 gardens throughout the city.

3. For more information, see http://thefoodproject.org/youth-programs.

4. Cerdá is often considered to be one of the first modern urban planners. For most of his career, he focused on the expansion of the city according to a utopian socialist vision. His 1857 plan for Barcelona remained the official city plan until 1953. It has been praised for accommodating the commercial, social, and cultural needs of its newly prosperous residents and for introducing a public-health centered vision of urban planning, but criticized for the destruction of the last well and many old buildings in Ciutat Vella as well as the exodus of middle-class residents away from the city center.

6   Advancing Broader Political Agendas

1. From "Local Administrations Facing Migratory Facts," signed in 1995 in Manlleu by the Federació de Municipis de Catalunya and the Associació Catalana de Municipis I Comarques.

References

Aalbers, Manuel, Wouter van Gent, and Fenne Pinkster. 2011. Comparing Deconcentrating Poverty Policies in the United States and the Netherlands: A Critical Reply to Stal and Zuberi. *Cities* 28 (3):260–264.

Abella, Martí. 2004. *Ciutat Vella: El centre històric reviscolat.* Barcelona: Edicions Universitat.

Abrahamson, Mark. 2004. *Global Cities.* New York: Oxford University Press.

Adger, W. N. 2006. Vulnerability. *Global Environmental Change* 16 (3):268–281.

Agyeman, Julian. 2002. Constructing Environmental (In)justice: Transatlantic Tales. *Environmental Politics* 11 (3):31–53.

Agyeman, Julian, Robert Bullard, and Bob Evans. 2003. *Just Sustainabilities: Development in an Unequal World.* Cambridge, Mass.: MIT Press.

Agyeman, Julian, and Tom Evans. 2003. Toward Just Sustainability in Urban Communities: Building Equity Rights with Sustainable Solutions. *Annals of the American Academy of Political and Social Science* 590:35–53.

Agyeman, Julian, and Yelena Ogneva-Himmelberger. 2009. *Environmental Justice and Sustainability in the Former Soviet Union.* Cambridge, Mass.: MIT Press.

Ahmad, Afroz. 1999. The Narmada Water Resources Project, India: Implementing Sustainable Development. *Ambio* 28 (5):398–403.

Ajuntament de Barcelona. 2005. *Projecte d'intervenció integral als barris de Santa Caterina i Sant Pere sector Casc Antic de la Ciutat Vella de Barcelona.* Barcelona: Ajuntament de Barcelona.

Ajuntament de Barcelona. 2009. *Memòria del projecte Allada Vermell i Pou de la Figuera reparació del paviment.* Barcelona: Ajuntament de Barcelona.

Ajuntament de Barcelona. 2010. El barri de Sant Pere, Santa Caterina i la Ribera estrena el CEM Parc de la Ciutadella. Barcelona: Ajuntament de Barcelona.

Alió, M. Angels, and Gerard Jori. 2010. La reforma ambiental de las ciudades: Visiones y propuestas del movimiento vecinal de Barcelona. *Scripta Nova: Revista Electrónica de Geografía y Ciencias Sociales* 14 (331):741–798.

Alkon, Alison Hope, and Julian Agyeman. 2011. *Cultivating Food Justice: Race, Class, and Sustainability, Food, Health, and the Environment.* Cambridge, Mass.: MIT Press.

Altman, Irwin, Setha Low, and Craig Maretzki. 1992. *Place Attachment.* New York: Plenum Press.

Anderson, Elijah. 1990. *Streetwise: Race, Class, and Change in an Urban Community.* Chicago: University of Chicago Press.

Anderson, Elijah. 2003. *A Place on the Corner.* 2nd ed. Chicago: University of Chicago Press.

Anguelovski, Isabelle, and Debra Roberts. 2011. Spatial (In)justice in Urban Climate Policy: A Multi-scale Analysis of Climate Impacts in Durban. In *Environmental Inequalities beyond Borders: Local Perspectives on Global Injustices,* ed. J. Carmin and J. Agyeman. Cambridge, Mass.: MIT Press.

Arbaci, Sonia, and Teresa Tapada-Berteli. 2012. Social Inequality and Urban Regeneration in Barcelona City Centre: Reconsidering Success. *European Urban and Regional Studies* 19 (3):287–311.

Bandy, J., and J. Smith. 2005. *Coalitions across Borders: Transnational Protest and the Neoliberal Order.* Lanham, Md.: Rowman & Littlefield.

Bartolomé Bárquez, Carlos, ed. 2004. *Historia de Centro Habana.* Havana: Ciudad de La Habana.

Bastos, Wanderley Rodrigues, Paulo Oliveira Gomes João, Ronaldo Cavalcante Oliveira, Ronaldo Almeida, Elisabete Lourdes Nascimento, Vicente Elias Bernardi José, Luiz Drude de Lacerda, Ene Glória da Silveira, and Wolfgang Christian Pfeiffer. 2006. Mercury in the Environment and Riverside Population in the Madeira River Basin, Amazon, Brazil. *Science of the Total Environment* 368 (1):344–351.

Bates, Robert. 1998. *Analytic Narratives.* Princeton, N.J.: Princeton University Press.

Beaulac, Julie, Elizabeth Kristjansson, and Steven Cummins. 2009. A Systematic Review of Food Deserts, 1966–2007. *Preventing Chronic Disease: Public Health Research, Practice, and Policy* 6 (3):A105.

Bell, Janice, Jeffrey Wilson, and Gilbert Liu. 2008. Neighborhood Greenness and Two-Year Changes in Body Mass Index of Children and Youth. *American Journal of Preventive Medicine* 35 (6):547–553.

Bickerstaff, Karen, Harriet Bulkeley, and J. O. E. Painter. 2009. Justice, Nature and the City. *International Journal of Urban and Regional Research* 33 (3):591–600.

Birch, Eugenie Ladner, and Susan M. Wachter. 2008. *Growing Greener Cities: Urban Sustainability in the Twenty-first Century.* Philadelphia: University of Pennsylvania Press.

Blodgett, Abigail. 2006. An Analysis of Pollution and Community Advocacy in "Cancer Alley": Setting an Example for the Environmental Justice Movement in St. James Parish, Louisiana. *Local Environment* 11 (6):647–661.

Blowfield, M., and J. G. Frynas. 2005. Setting New Agendas: Critical Perspectives on Corporate Social Responsibility in the Developing World. *International Affairs* 81 (3):499–513.

Blum, Elizabeth D. 2008. *Love Canal Revisited: Race, Class, and Gender in Environmental Activism.* Lawrence: University Press of Kansas.

Blum, Terry C., and Paul W. Kingston. 1984. Homeownership and Social Attachment. *Sociological Perspectives* 27:159–180.

Boelens, R., L. Cremers, and M. Zwarteveen. 2011. *Justicia Hídrica: Acumulación, Conflicto y Acción Social.* Lima: Instituto de Estudios Peruanos.

Bondi, Liz. 1993. Locating Identity Politics. In *Place and the Politics of Identity*, ed. M. Keith and S. Pile. London: Routledge.

Bonet, Mariano, and Pedro Mas. 1996. National and Provincial Risk Factor Survey for Noncommunicable Diseases. Havana: INHEM/ONE/MINSAP.

Borja, Jordi. 2004. Barcelona y su urbanismo: Exitos pasados, desafios presentes, oportunidades futuras. In *Urbanismo en el siglo XXI: Una visión crítica—Bilbao, Madrid, Valencia, Barcelona*, ed. J. Borja, Z. Muxí and J. Cenicacelaya. Barcelona: Escola Tècnica Superior d'Arquitectura de Barcelona, Edicions UPC.

Borja, Jordi, Manuel Castells, Mireia Belil, and Chris Benner. 1997. *Local and Global: The Management of Cities in the Information Age.* London: Earthscan.

Borja, Jordi, Zaida Muxí, and Javier Cenicacelaya. 2004. *Urbanismo en el siglo XXI: Una visión crítica—Bilbao, Madrid, Valencia, Barcelona.* Barcelona: Escola Tècnica Superior d'Arquitectura de Barcelona, Edicions UPC.

Boston Parks and Recreation Department. 2007. *Dennis Street Park: Developing an Urban Village Common.* Boston: City of Boston.

Boston Public Health Commission. 2004. *Health Status Report for Roxbury.* Boston: Boston Public Health Commission.

Boston Schoolyard Initiative (BSI). 2009. Boston Schoolyard Initiative. Boston: BSI.

Brady, Henry, and David Collier. 2004. *Rethinking Social Inquiry: Diverse Tools, Shared Standards.* Lanham, Md.: Rowman & Littlefield.

Brenner, Neil. 2009. Cities and Territorial Competitiveness. In *The Sage Handbook of European Studies*, ed. C. Rumford. London: Sage.

Brown, Barbara, and Douglas Perkins. 1992. *Disruptions in Place Attachment*, ed. I. Altman and S. Low. New York: Springer.

Brown, Barbara, Douglas Perkins, and Graham Brown. 2004. Erratum to Place Attachment in a Revitalizing Neighborhood: Individual and Block Levels of Analysis. *Journal of Environmental Psychology* 24 (1):141.

Brulle, Robert, and David Pellow. 2006. Environmental Justice: Human Health and Environmental Inequalities. *Annual Review of Public Health* 27 (1):103–124.

Bryant, Bunyan I., and Paul Mohai. 1992. *Race and the Incidence of Environmental Hazards: A Time for Discourse.* Boulder, Colo.: Westview Press.

Brysk, Alison. 2000. *From Tribal Village to Global Village: Indian Rights and International Relations in Latin America*. Stanford, Calif.: Stanford University Press.

Bullard, Robert. 1990. *Dumping in Dixie: Race, Class, and Environmental Quality*. Boulder, Colo.: Westview Press.

Bullard, Robert. 2005. *The Quest for Environmental Justice: Human Rights and the Politics of Pollution*. San Francisco: Sierra Club Books.

Bullard, Robert D. 2007. *Growing Smarter: Achieving Livable Communities, Environmental Justice, and Regional Equity*. Cambridge, Mass.: MIT Press.

Büthe, Tim. 2002. Taking Temporality Seriously: Modeling History and the Use of Narratives as Evidence. *American Political Science Review* 96 (3):481–493.

Calavita, Nico, and Amador Ferrer. 2004. Behind Barcelona's Success Story: Citizen Movements and Planners' Power. In *Transforming Barcelona*, ed. T. Marshall. London: Routledge.

Čaldarović, Ognjen, and Jana Šarinić. 2009. First Signs of Gentrification? Urban Regeneration in the Transitional Society: The Case of Croatia. *Sociologija i prostor* 46 (3–4):369–381.

Campbell, S., and S. S. Fainstein. 2003. *Readings in planning theory*. 2nd ed. Malden, Mass.: Blackwell Publishers.

Capel, Horacio. 2005. *El Modelo Barcelona: Un examen crítico*. Santiago: Eure.

Capel, Horacio. 2007. El debate sobre la construcción de la ciudad y el llamado "Modelo Barcelona." *Scripta Nova: Revista Electrónica de Geografía y Ciencias Sociales* 11 (233):741–798.

Caridad, Maria, and Roberto Medina. 2001. Agricultura y ciudad: Una clave para la sustentabilidad. Havana: Fundación Antonio Núñez Jiménez de la Naturaleza el Hombre.

Carmin, JoAnn, and Julian Agyeman, eds. 2011. *Environmental Inequalities beyond Borders: Local Perspectives on Global Injustices*. Cambridge, Mass.: MIT Press.

Carruthers, David. 2008. *EnvironmentalJjustice in Latin America: Problems, Promise, and Practice*. Cambridge, Mass.: MIT Press.

Casademunt, Àlex, Eva Alfama, Gerard Coll-Planas, Helena Cruz, and Marc Martí. 2007. *Per una nova cultura del territori? Mobilitzacions i conflictes territorials*. Barcelona: Icària.

Castells, Manuel. 1977. *The Urban Question: A Marxist Approach*. Cambridge, Mass.: MIT Press.

Castells, Manuel. 1983. *The City and the Grassroots: A Cross-Cultural Theory of Urban Social Movements*. London: Arnold.

Chambers, Stefanie. 2007. Minority Empowerment and Environmental Justice. *Urban Affairs Review* 43 (1):28–54.

Charmaz, Kathy. 2006. *Constructing Grounded Theory: A Practical Guide through Qualitative Analysis*. London: Sage.

Chatterton, Paul. 2010. Seeking the Urban Common: Furthering the Debate on Spatial Justice. *City* 14 (6):625–628.

Chavis, David, and Abraham Wandersman. 1990. Sense of Community in the Urban Environment: A Catalyst for Participation and Community Development. *American Journal of Community Psychology* 18 (1):55–81.

Checkel, Jeffrey. 2005. Process Tracing. In *Qualitative Methods in International Relations: A Pluralist Guide,* ed. Audie Klotz and Deepa Prakash. Houndmills. UK: Palgrave MacMillan.

Checker, Melissa. 2008. *Polluted Promises: Environmental Racism and the Search for Justice in a Southern Town.* New York: New York University Press.

Checker, Melissa. 2011. Wiped Out by the "Greenwave": Environmental Gentrification and the Paradoxical Politics of Urban Sustainability. *City & Society* 23 (2):210–229.

Chion, Miriam. 2009. Producing Urban Vitality: The Case of Dance in San Francisco. *Urban Geography* 30 (4):416–439.

Clark, Kenneth. 1989. *Dark Ghetto: Dilemmas of Social Power.* Middletown, Conn.: Wesleyan University Press.

Cohrun, Steven. 1994. Understanding and Enhancing Neighborhood Sense of Community. *Journal of Planning Literature* 9:92–99.

Comité Estatal de Estadísticas. 1999. *Anuario Estadístico de Cuba, 1988.* Havana: Comité Estatal de Estadísticas.

Corburn, Jason. 2005. *Street Science: Community Knowledge and Environmental Health Justice.* Cambridge, Mass.: MIT Press.

Corburn, Jason. 2009. *Toward the Healthy City: People, Places, and the Politics of Urban Planning.* Cambridge, Mass.: MIT Press.

Corcoran, Mary. 2002. Place Attachment and Community Sentiment in Marginalized Neighbourhoods: A European Case Study. *Canadian Journal of Urban Research* 11 (1):47–68.

Cotton, Jeremiah. 1992. A Reexamination of How We Measure Poverty: A Profile of Black Poverty in Boston. In *Perspectives on Poverty in Boston's Black Community,* ed. J. Jennings and M. E. Carrion. Boston: Boston Persistent Poverty Project, Boston Foundation.

Cox, Kevin. 1982. Housing Tenure and Neighborhood Activism. *Urban Affairs Quarterly* 18:107–129.

Coyula, Mario. 2008. Los Talleres de Transformación Integral del Barrio en La Habana: Una experiencia de planeamiento alternativo a nivel de base. Paper delivered at a conference on Ciudades en (re)construcción: Necesidades sociales, transformación y mejora de barrios, Centre de Cultura Contemporània, Barcelona, December 12–14.

Coyula, Mario. 2009a. Los muchos centros de La Habana. Paper presented at The New School, New York, August 24.

Coyula, Mario. 2009b. La toma de la grand ciutad blanca. Galeriascubanas.com, October 16.

Coyula, Mario, and Jill Hamberg. 2003. The Case of Havana, Cuba. In *Understanding Slums: Case Studies for Global Report on Human Settlements*. London: Development Planning Unit, University College London, and Nairobi: United Nations Human Settlement Program.

Cresswell, Tim. 2009. Place. In *International Encyclopedia of Human Geography*, ed. R. Kitchin and N. Thrift. Oxford: Elsevier.

Crump, Jeff. 2003. The End of Public Housing as We Know It: Public Housing Policy, Labor Regulation and the U.S. City. *International Journal of Urban and Regional Research* 27 (1):179–187.

Cummins, Steven, Sarah Curtis, Ana V. Diez-Roux, and Sally Macintyre. 2007. Understanding and Representing "Place" in Health Research: A Relational Approach. *Social Science & Medicine* 65 (9):1825–1838.

Cummins, Steven, and Sally Macintyre. 2002. "Food Deserts": Evidence and Assumption in Health Policy Making. *British Medical Journal* 325 (7361):436–438.

Curran, Winifred, and Trina Hamilton. 2012. Just Green Enough: Contesting Environmental Gentrification in Greenpoint, Brooklyn. *Local Environment* 17 (9):1027–1042.

Curtis, Russell, and Louis Zurcher. 1973. Stable Resources of Protest Movements: The Multi-Organizational Field. *Social Forces* 52 (1):53–61.

Davey, Iain. 2009. Environmentalism of the Poor and Sustainable Development: An Appraisal. *JOAAG* 4 (1):1–10.

Davidson, William, and Patrick Cotter. 1993. Psychological Sense of Community and Support for Public School Taxes. *American Journal of Community Psychology* 21 (1):59–66.

Davis, Diane, and Christina Rosan. 2004. Social Movements in the Mexico City Airport Controversy: Globalization, Democracy, and the Power of Distance. *Mobilization: An International Quarterly* 9 (3):279–293.

Davis, Jonathan. 1991. *Contested Ground: Collective Action and the Urban Neighborhood*. Ithaca, N.Y.: Cornell University Press.

De Certeau, Michel. 1984. *The Practice of Everyday Life*. Berkeley: University of California Press.

De la Fuente, Alejandro. 2000. *Una nación para todos: Raza, desigualdad y política en Cuba, 1900–2000*. Madrid: Colibrí.

Delgado, Manuel. 2007. *La ciudad mentirosa: Fraude y miseria del modelo Barcelona*. Barcelona: Libros de la Catarata.

Della Porta, Donatella, and Sidney Tarrow. 2005. *Transnational Protest and Global Activism: People, Passions, and Power*. Lanham, Md.: Rowman & Littlefield.

de Souza, Marcelo Lopes. 2006. Social Movements as "Critical Urban Planning" Agents. *City* 10 (3):327–342.

Dewar, Margaret, and Robert Linn. Forthcoming. Remaking Brightmoor. In *Mapping Detroit*, ed. J. T. Manning and H. Bekkering. Detroit: Wayne State University Press.

Diaz, David. 2005. *Barrio Urbanism: Chicanos, Planning, and American Cities.* New York: Routledge.

Díaz, Joel. 2001. *Cayo Hueso: Un proceso de mejoramiento en un contexto urbano, participación y protagonismo de los actores Comunitarios.* Washington, D.C.: Latin American Studies Association.

Diaz Veizades, Jeannette, and Edward Chang. 1996. Building Cross-Cultural Coalitions: A Case-Study of the Black-Korean Alliance and the Latino-Black Roundtable. *Ethnic and Racial Studies* 19:680–700.

Di Chiro, Giovanna. 2008. Living Environmentalisms: Coalition Politics, Social Reproduction, and Environmental Justice. *Environmental Politics* 17 (2):276–298.

Dikeç, Mustafa. 2001. Justice and the Spatial Imagination. *Environment & Planning A* 33 (10):1785–1805.

Dilla, Haroldo, and Philip Oxhorn. 2002. The Virtues and Misfortunes of Civil Society in Cuba. *Latin American Perspectives* 29 (4):11–30.

Dilla Alfonso, Haroldo, Armando Fernández Soriano, and Margarita Castro Flores. 1998. Movimientos barriales en Cuba: Un análisis comparativo. In *Participación social: Desarrollo urbano y comunitario,* ed. Aurora Vazquez and Roberto Dávalos. Havana: Universidad de la Habana.

Dobson, Andrew. 1998. *Justice and the Environment: Conceptions of Environmental Sustainability and Theories of Distributive Justice.* Oxford: Oxford University Press.

Dooling, Sarah. 2009. Ecological Gentrification: A Research Agenda Exploring Justice in the City. *International Journal of Urban and Regional Research* 33 (3):621–639.

Doolittle, Robert, and Donald MacDonald. 1978. Communication and Sense of Community in a Metropolitan Neighborhood: A Factor Analytic Examination. *Communication Quarterly* 26 (1):1–27.

Dorchester Bay Economic Development Corporation. 2011. *Current Projects.* http://www.dbedc.org/affordablehousing/projects/currentprojects.html., accessed March 29, 2011.

Downey, Liam, and Brian Hawkins. 2008. Race, Income, and Environmental Inequality in the United States. *Sociological Perspectives* 51 (4):759–781.

Doyle, Timothy, and Melissa Risely. 2008. *Crucible for Survival: Environmental Security and Justice in the Indian Ocean Region.* New Brunswick, N.J.: Rutgers University Press.

Dudley Street Neighborhood Initiative (DSNI). 2008. *History.* http://www.dsni.org/history.shtml, accessed March 29, 2011.

Dunn, Richard A. 2010. The Effect of Fast-Food Availability on Obesity: An Analysis by Gender, Race, and Residential Location. *American Journal of Agricultural Economics* 92 (4):1149–1164.

Eckstein, Susan. 2003. *Back from the Future: Cuba under Castro.* 2nd ed. New York: Routledge.

Espina, Mayra Prieto, Lilia Núñez Moreno, Lucy Martín Posada, Laritza Vega Quintana, Adrián Rodríguez Chailloux, and Gisela Ángel Sierra. 2005. Heterogenización y desigualdades en la ciudad: Diagnóstico y perspectivas. *Boletín Electrónico del CIPS* 1 (4):1–18.

Estabrooks, Paul, Rebecca Lee, and Nancy Gyurcsik. 2003. Resources for Physical Activity Participation: Does Availability and Accessibility Differ by Neighborhood Socioeconomic Status? *Annals of Behavioral Medicine* 25 (2):100–104.

Evans, Geoff, James Goodman, and Nina Lansbury. 2002. *Moving Mountains: Communities Confront Mining and Globalisation*. London: Zed Books.

Evans, P. 2002. *Livable Cities? Urban Struggles for Livelihood and Sustainability*. Berkeley: University of California Press.

Fagotto, Elena, and Archon Fung. 2006. Empowered Participation in Urban Governance: The Minneapolis Neighborhood Revitalization Program. *International Journal of Urban and Regional Research* 30 (3):638–655.

Fainstein, Susan. 1999. Can We Make the Cities We Want? In *The Urban Moment: Cosmopolitan Essays on the Late Twentieth-Century City*, ed. R. A. Beauregard and S. Body-Gendrot. Thousand Oaks, Calif.: Sage Publications.

Fainstein, Susan. 2006. Planning and the Just City. Paper presented at the Searching for the Just City conference, Columbia University, April 29.

Fainstein, Susan S. 2011. *The Just City*. Ithaca, N.Y.: Cornell University Press.

Falk, William. 2004. *Rooted in Place: Family and Belonging in a Southern Black Community*. New Brunswick, N.J.: Rutgers University Press.

Food and Agriculture Organization (FAO). 2001. *Food Balance Sheets*. Rome: United Nations Food and Agriculture Organization.

Ferriol Muruaga, Angela, ed. 1998. *La seguridad alimentaria en Cuba, Cuba: Crisis, ajuste y situación social 1990–96*. Havana: Editorial de Ciencias Sociales.

Fisher, Robert. 1984. *Let the People Decide: Neighborhood Organizing in America*. Boston: Twayne.

Fitzgerald, Joan. 2010. *Emerald Cities: Urban Sustainability and Economic Development*. New York: Oxford University Press.

Flora, Cornelia, and Jan Flora. 2006. Creating Social Capital. In *Rooted in the Land: Essays on Community and Lace*, ed. W. Vitek and W. Jackson. New Haven, Conn.: Yale University Press.

Francis, Mark, and Randolph Hester, eds. 1999. *The Meaning of Gardens*. Cambridge, Mass.: MIT Press.

Fredrickson, Laura, and Dorothy Anderson. 1999. A Qualitative Exploration of the Wilderness Experience as a Source of Spiritual Inspiration. *Journal of Environmental Psychology* 19:21–40.

Frieden, Thomas R. 2010. A Framework for Public Health Action: The Health Impact Pyramid. *American Journal of Public Health* 100 (4):590–595.

Friedmann, John. 1986. The World City Hypothesis. *Development and Change* 17 (1):69–83.

Friedmann, John. 2000. The Good City: In Defense of Utopian Thinking. *International Journal of Urban and Regional Research* 24 (2):460–472.

Frynas, J. G. 2005. The False Developmental Promise of Corporate Social Responsibility: Evidence from Multinational Oil Companies. *International Affairs* 81 (3):581–598.

Fullilove, Mindy. 1996. Psychiatric Implications of Displacement: Contributions from the Psychology of Place. *American Journal of Psychiatry* 153 (12):1516–1523.

Fullilove, Mindy. 2004. *Root Shock: How Tearing Up City Neighborhoods Hurts America, and What We Can Do about It*. New York: One World / Ballantine Books.

Gamper-Rabindran, Shanti, Ralph Mastromonaco, and Christopher Timmins. 2011. Valuing the Benefits of Superfund Site Remediation: Three Approaches to Measuring Localized Externalities. NBER Working Paper No. 16655. National Bureau of Economic Research.

Gamper-Rabindran, Shanti, and Christopher Timmins. 2011. Hazardous Waste Cleanup, Neighborhood Gentrification, and Environmental Justice: Evidence from Restricted Access Census Block Data. *American Economic Review* 101 (3):620–624.

Gamson, William. 1990. *The Strategy of Social Protest*. 2nd ed. Belmont, Calif.: Wadsworth.

Gans, Herbert. 1962. *The Urban Villagers: Group and Class in the Life of Italian-Americans*. New York: Free Press of Glencoe.

Ganz, Marshall. 2000. Resources and Resourcefulness: Strategic Capacity in the Unionization of California Agriculture, 1959–1966. *American Journal of Sociology* 105 (4):1003–1062.

Garbin, David, and Gareth Millington. 2012. Territorial Stigma and the Politics of Resistance in a Parisian *Banlieue*: La Courneuve and Beyond. *Urban Studies* (Edinburgh, Scotland) 49 (10):2067–2083.

Garfield, Richard, and Sarah Santana. 1997. The Impact of the Economic Crisis and the U.S. Embargo on Health in Cuba. *American Journal of Public Health* 87 (1):15–20.

Gauna, Eileen. 2008. El Dia de los Muertos: The Death and Rebirth of the Environmental Movement. *Environmental Law* (Northwestern School of Law) 38 (2):457–472.

Generalitat de Catalunya and Ajuntament de Barcelona. 2006. *Programa de millora de barris: Santa Caterina i Sant Pere*. Barcelona: Generalitat de Catalunya and Ajuntament de Barcelona.

George, Alexander L., and Andrew Bennett. 2005. *Case Studies and Theory Development in the Social Sciences*. Cambridge, Mass.: MIT Press.

George, Alexander, and Tim McKeown. 1985. Case Studies and Theories of Organizational Decision Making. In *Advances in Information*, ed. Robert Coulam and Richard Smith, 43–68. Greenwich, Conn.: JAI Press.

Gerlach-Spriggs, Nancy, Richard Kaufman, and Sam Bass Warner. 2004. *Restorative Gardens: The Healing Landscape.* New Haven, Conn.: Yale University Press.

Gerring, John. 2007. *Case Study Research: Principles and Practices.* New York: Cambridge University Press.

Gieryn, Thomas. 2000. A Space for Place in Sociology. *Annual Review of Sociology* 26:463–496.

Glaser, Barney, and Anselm Strauss. 1967. *The Discovery of Grounded Theory: Strategies for Qualitative Research.* Chicago.: Aldine.

Goetz, Edward. 2003. *Clearing the Way: Deconcentrating the Poor in Urban America.* Washington, D.C.: Urban Institute Press.

Goldman, Michael, and Wesley Longhofer. 2009. Making World Cities. *Contexts* 8 (1):32–36.

Gómez, Indira Brito, and Jetter González Nieda. 2005. *Detección de valores en el Municipio Centro Habana: Consejos Populares Cayo Hueso.* Dragones y Los Sitios. La Habana: Instituto Superior Politécnico José Antonio Echevarria, CUJAE.

Gotham, Kevin. 1999. Political Opportunity, Community Identity, and the Emergence of Local Anti-expressway Movement. *Social Problems* 46 (3):332–354.

Gotham, Kevin. 2003. Toward an Understanding of the Spatiality of Urban Poverty: The Urban Poor as Spatial Actors. *International Journal of Urban and Regional Research* 27 (3):723–737.

Gotham, Kevin, and Krista Brumley. 2002. Using Space: Agency and Identity in a Public-Housing Development. *City & Community* 1 (3):267–289.

Gottardi, Mario. 2007. Quitar los bichos: O sea, echar un hombre a la calle. Trabajos de alumnos - Prensa, Master en periodismo BCNY, Institute for Life-Long Learning, University of Barcelona.

Gottlieb, R., and A. Joshi. 2010. *Food Justice.* Cambridge, Mass.: MIT Press.

Gottlieb, Robert. 2005. *Forcing the Spring: The Transformation of the American Environmental Movement.* Rev. and updated ed. Washington, D.C.: Island Press.

Gottlieb, Robert. 2009. Where We Live, Work, Play . . . and Eat: Expanding the Environmental Justice Agenda. *Environmental Justice* 2 (1):7–8.

Gould, Kenneth, and Tammy Lewis. 2012. The Environmental Injustice of Green Gentrification: The Case of Brooklyn's Prospect Park. In *The World in Brooklyn: Gentrification, Immigration, and Ethnic Politics in a Global City*, ed. J. DeSena and T. Shortell. Lanham, Md.: Lexington Books.

Gould, Kenneth, Tammy Lewis, and J. Timmons Roberts. 2004. Blue-Green Coalitions: Constraints and Possibilities in the Post 9-11 Political Environment. *Journal of World-System Research* 10 (1):90–116.

Green, James. 1986. The Making of Mel King's Rainbow Coalition: Political Changes in Boston 1963–1983. In *From Access to Power: Black Politics in Boston*, ed. J. Jennings and M. King. Cambridge, Mass.: Schenkman.

Green Map. 2007. Mapa Verde Cuba: Rural and Urban Green Mapping in Diverse Provinces across Cuba. Green Map, Havana.

Gregory, Steven. 1998. *Black Corona: Race and the Politics of Place in an Urban Community*. Princeton, N.J.: Princeton University Press.

Guha, Ramachandra. 1989. *The Unquiet Woods: Ecological Change and Peasant Resistance in the Himalaya*. Ranikhet, India: Permanent Black.

Guha, Ramachandra. 2000. *Environmentalism: A Global History*. New York: Longman.

Guy, Cliff, Clarke Graham, and Eyre Heather. 2004. Food Retail Change and the Growth of Food Deserts: A Case Study of Cardiff. *International Journal of Retail & Distribution Management* 32 (2–3):72–88.

Hache, Alexandra. 2005. Ville-Monde Barcelone: Projets urbains globaux et revendications territoriales. *Ville-Socio-Anthropologie* http://socio-anthropologie. revues.org/433.

Hagerman, Chris. 2007. Shaping Neighborhoods and Nature: Urban Political Ecologies of Urban Waterfront Transformations in Portland, Oregon. *Cities* 24 (4):285–297.

Harrison, Jill Lindsey. 2011. *Pesticide Drift and the Pursuit of Environmental Justice: Food, Health, and the Environment*. Cambridge, Mass.: MIT Press.

Harvey, David. 1981. The Urban Process under Capitalism: A Framework for Analysis. In *Urbanization and Urban Planning in Capitalist Society*, ed. M. J. Dear and A. J. Scott. London: Methuen.

Harvey, David. 1989. *The Conditions of Postmodernity: An Enquiry into the Origins of Cultural Change*. Oxford: Blackwell.

Harvey, David. 1996. *Justice, Nature and the Geography of Difference*. Cambridge, Mass.: Blackwell.

Harvey, David. 2003. The Right to the City. *International Journal of Urban and Regional Research* 27 (4):939–941.

Harvey, David. 2012. *Rebel Cities: From the Right to the City to the Urban Revolution*. London: Verso.

Hastings, Annette. 2007. Territorial Justice and Neighbourhood Environmental Services: A Comparison of Provision to Deprived and Better-off Neighbourhoods in the UK. *Environment and Planning* C (25):896–917.

Hayden, Dolores. 1995. *The Power of Place: Urban Landscapes as Public History*. Cambridge, Mass.: MIT Press.

Haymes, Stephen Nathan. 1995. *Race, Culture, and the City: A Pedagogy for Black Urban Struggle*. Albany: SUNY Press.

Hearn, Adrian. 2008. *Cuba: Religion, Social Capital, and Development*. Durham, N.C.: Duke University Press.

Hernández, Rafael, Aurora Vásquez, Rubén Zardoya, and Miguel Mejiles. 2004. Qué significa ser marginal? In *Ultimo Jueves: Los Debates de Tema* 1, 83–130. Havana: Ediciones Unión.

Hess, D. J. 2009. *Localist Movements in a Global Economy: Sustainability, Justice, and Urban Development in the United States*. Cambridge, Mass.: MIT Press.

Heynen, Nik, Harold Perkins, and Parama Roy. 2006. The Political Ecology of Uneven Urban Green Space. *Urban Affairs Review* 42 (1):3–25.

Hillier, Amy. 2003. Redlining and the Home Owners' Loan Corporation. *Journal of Urban History* 29 (4):394–420.

Hillier, Amy, Carolyn Cannuscio, Allison Karpyn, Jacqueline McLaughlin, Mariana Chilton, and Karen Glanz. 2011. How Far Do Low-Income Parents Travel to Shop for Food? Empirical Evidence from Two Urban Neighborhoods. *Urban Geography* 32 (5):712–729.

Hilson, Gavin. 2002. An Overview of Land Use Conflicts in Mining Communities. *Land Use Policy* 19 (1):65–73.

Hou, Jeffrey. 2010. *Insurgent Public Space: Guerrilla Urbanism and the Remaking of Contemporary Cities*. New York: Routledge.

Hummon, David Mark. 1992. Community Attachment: Local Sentiment and Sense of Place. In *Place Attachment*, ed. I. Altman and S. M. Low. New York: Plenum Press.

Hunter, Albert, and Suzanne Staggenborg. 1986. Communities Do Act: Neighborhood Characteristics, Resource Mobilization, and Political Action by Local Community Organizations. *Social Science Journal* 23 (2):169–180.

Idealista. 2007. *El alquiler sube un 4,4% en barcelona y un 6,1% en madrid durante el primer semestre*. Madrid: Idealista.

Inam, Aseem. 2005. *Planning for the Unplanned: Recovering from Crises in Megacities*. London: Routledge.

Instituto de Planificación Física de Cuba. 2002. *Plan de rehabilitaciónu del Municipio Centro Habana*. Havana: Instituto de Planficación Física de Cuba.

Instituto Nacional de Higiene, Epidemiología y Microbiología. 1999. *Plan de medidas ejecutadas por el Gobierno y la Comunidad (PMGC) en Cayo Hueso, 1996–1999*. Havana: Instituto Nacional de Higiene, Epidemiología y Microbiología.

Kearns, Ade. 2002. Response: From Residential Disadvantage to Opportunity? Reflections on British and European Policy and Research. *Housing Studies* 17 (1):145–150.

Keck, M. E., and K. Sikkink. 1998. *Activists beyond Borders: Advocacy Networks in International Politics*. Ithaca, N.Y.: Cornell University Press.

Kefalas, Maria. 2003. *Working-Class Heroes: Protecting Home, Community, and Nation in a Chicago Neighborhood*. Berkeley: University of California Press.

King, Mel. 1981. *Chain of Change: Struggles for Black Community Development*. Boston: South End Press.

Kirkpatrick, Anthony. 1997. The U.S. Attack on Cuba's Health. *Canadian Medical Association Journal* 157 (3):281–284.

Kitschelt, H. P. 1986. Political Opportunity Structures and Political Protest: Anti-Nuclear Movements in Four Democracies. *British Journal of Political Science* 16:57–85.

Klandermans, Bert. 1992. The Social Construction of Protests and Multiorganizational Fields. In *Frontiers in Social Movement Theory*, ed. A. D. Morris and C. M. Mueller. New Haven, Conn.: Yale University Press.

Kuo, Frances, and William Sullivan. 2001. Environment and Crime in the Inner City. *Environment and Behavior* 33 (3):343–367.

Kuo, Frances, William Sullivan, Rebekah Coley, and Liesette Brunson. 1998. Fertile Ground for Community: Inner-City Neighborhood Common spaces. *American Journal of Community Psychology* 26 (6):823–851.

Kurtz, Hilda. 2007. Gender and Environmental Justice in Louisiana: Blurring the Boundaries of Public and Private Spheres. *Gender, Place and Culture* 14 (4):409–426.

Lahuerta, Juan José. 2005. *Destrucción de Barcelona*. Barcelona: Mudito.

Lamont, Michelle, and Virag Molnar. 2002. The Study of Boundaries in the Social Sciences. *Annual Review of Sociology* 28:167–196.

Landry, Shawn, and Jayajit Chakraborty. 2009. Street Trees and Equity: Evaluating the Spatial Distribution of an Urban Amenity. *Environment & Planning A* 41 (11):2651–2670.

Lawson, Laura. 2005. *City Bountiful: A Century of Community Gardening in America*. Berkeley: University of California Press.

Layzer, Judith. 2006. *The Environmental Case: Translating Values into Policy*. 2nd ed. Washington, D.C.: CQ Press.

Lefebvre, Henri. 1968. *Le Droit à la ville: Société et urbanisme*. Paris: Anthropos.

Lefebvre, Henri. 1972. *Espace et politique: Le droit à la ville II*. Paris: Éditions Anthropos.

Lefebvre, Henri, Eleonore Kofman, and Elizabeth Lebas. 1996. *Writings on Cities*. Oxford: Blackwell.

Leiva-Vojkovic, Enrique, Ivan Miró, and Xavier Urbano. 2007. *De la protesta al contrapoder: Nous protagonismes socials en la Barcelona metropolitana*. Barcelona: Virus.

Lerner, S. 2005. *Sacrifice Zones: The Front Lines of Toxic Chemical Exposure in the United States*. Cambridge, Mass.: MIT Press.

Lévi-Strauss, Claude. 1962. *La pensée sauvage*. Paris: Plon.

Levine, Hillel, and Lawrence Harmon. 1992. *The Death of an American Jewish Community: A Tragedy of Good Intentions*. New York: Free Press.

Logan, John R., and Harvey Molotch. 1987. *Urban Fortunes: The Political Economy of Place*. Berkeley: University of California Press.

Loh, Penn, and Phoebe Eng. 2010. *Environmental Justice and the Green Economy: A Vision Statement and Case Studies for Just and Sustainable Solutions*. Boston: Alternatives for Community and the Environment.

Loh, Penn, and Jodi Sugerman-Brozan. 2002. Environmental Justice Organizing for Environmental Health: Case Study on Asthma and Diesel Exhaust in Roxbury,

Massachusetts. *Annals of the American Academy of Political and Social Science* 584 (1):110–124.

Lopez, Russ, Richard Campbell, and James Jennings. 2008. The Boston Schoolyard Initiative: A Public-Private Partnership for Rebuilding Urban Play Spaces. *Journal of Health Politics, Policy and Law* 33 (3):617–638.

Lovasi, Gina, Malo Hutson, Monica Guerra, and Kathryn Neckerman. 2009. Built Environments and Obesity in Disadvantaged Populations. *Epidemiologic Reviews* 31 (1):7–20.

Low, Setha, and Irwin Altman. 1992. Place Attachment: A Conceptual Inquiry. In *Place Attachment*, ed. Irwin Altman and Setha M. Low. New York: Plenum.

Low, Setha M., and Denise Lawrence-Zúñiga. 2003. *The Anthropology of Space and Place: Locating Culture*. Malden, Mass.: Blackwell.

Lowrey, Kathleen. 2008. Incommensurability and New Economic Strategies among Indigenous and Traditional Peoples. *Journal of Political Ecology* 15:61–74.

Lucas, Karen, ed. 2004. *Running on Empty: Transport, Social Exclusion and Environmental Justice*. Bristol, UK: Policy Press.

Lukas, Anthony. 1985. *Common Ground: A Turbulent Decade in the Lives of Three American Families*. New York: Knopf.

Maantay, Juliana. 2002. Zoning Law, Health, and Environmental Justice: What's the Connection? *Journal of Law, Medicine & Ethics* 30 (4):572–593.

Maantay, Juliana. 2007. Asthma and Air Pollution in the Bronx: Methodological and Data Considerations in Using GIS for Environmental Justice and Health Research. *Health & Place* 13 (1):32.

MacDonald, Michael. 1999. *All Souls: A Family Story from Southie*. Boston, Mass.: Beacon Press.

Maeckelbergh, Marianne. 2012. Mobilizing to Stay Put: Housing Struggles in New York City. *International Journal of Urban and Regional Research* 36 (4):655–673.

Maller, Cecily, Mardie Townsend, Anita Pryor, Peter Brown, and Lawrence St. Leger. 2006. Healthy Nature—Healthy People: "Contact with Nature" as an Upstream Health Promotion Intervention for Populations. *Health Promotion International* 21 (1):45–54.

Manzo, L.C., and Douglas Perkins. 2006. Finding Common Ground: The Importance of Place Attachment to Community Participation and Planning. *Journal of Planning Literature* 20 (4):335.

Manzo, Lynne. 2003. Beyond House and Haven: Toward a Revisioning of Emotional Relationships with Places. *Journal of Environmental Psychology* 23:47–61.

Manzo, Lynne, Rachel Kleit, and Dawn Couch. 2008. Moving Three Times Is Like Having Your House on Fire Once: The Experience of Place and Impending Displacement among Public Housing Residents. *Urban Studies* 45 (9):1855–1878.

Marcus, Clare, and Marni Barnes. 1999. *Healing Gardens: Therapeutic Benefits and Design Recommendations*. New York: Wiley.

Marcuse, Peter. 2009a. From Critical Urban Theory to the Right to the City. *City* 13 (2–3):185–197.

Marcuse, Peter. 2009b. *Searching for the Just City: Debates in Urban Theory and Practice*. London: Routledge.

Marcuse, Peter. 2009c. Spatial Justice: Derivative but Causal of Social Injustice. *Spatial Justice* 1:49–57.

Marshall, Tim. 2004. *Transforming Barcelona*. London: Routledge.

Martín, Anna. 2007. *Diagnòstic Socioeconòmic i Ambiental Del Casc Antic*. Barcelona: Pla Integral Casc Antic.

Martínez-Alier, Joan. 1991. Ecology and the Poor: A Neglected Dimension of Latin American History. *Journal of Latin American Studies* 23:621–639.

Martinez-Alier, Joan. 2001. Mining Conflicts, Environmental Justice, and Evaluation. *Journal of Hazardous Materials* 86 (1–3):153–170.

Martínez-Alier, Joan. 2002. *The Environmentalism of the Poor: A Study of Ecological Conflicts and Valuation*. Cheltenham, UK: Edward Elgar.

Martinez-Alier, Joan. 2009. Social Metabolism, Ecological Distribution Conflicts, and Languages of Valuation. *Capitalism, Nature, Socialism* 20 (1):58–87.

Mas, Maria, and Toni Verger. 2004. Un forat de la vergonya al Casc Antic de Barcelona. In *Barcelona, marca registrada: Un model per desarmar*, ed. E. Unió Temporal. Barcelona: Virus.

Massachusetts Division of Health Care Finance and Policy (DHCFP). 1997. Untitled study. Commonwealth of Massachusetts, Boston.

Massey, Doreen. 1994. *Space, Place, and Gender*. Minneapolis: University of Minnesota Press.

Massey, Douglas S., and Nancy A. Denton. 1993. *American Apartheid: Segregation and the Making of the Underclass*. Cambridge, Mass.: Harvard University Press.

May, Reuben A. Buford. 2001. *Talking at Trena's: Everyday Conversations at an African American Tavern*. New York: New York University Press.

Mayer, Margit. 2009. The "Right to the City" in the Context of Shifting Mottos of Urban Social Movements. *City* 13 (2–3):362–374.

Mayer, Margit. 2012. Moving beyond "Cities for People, Not for Profit." *City* 16 (4):481–483.

McAdam, Doug. 1982. *Political Process and the Development of Black Insurgency, 1930–1970*. Chicago: University of Chicago Press.

McAuley, William. 1998. History, Race and Attachment to Place among Elders in the Rural All-Black Towns of Oklahoma. *Journal of Gerontology* 53B:S35–S45.

McCammon, Holly, and Karen Campbell. 2002. Allies on the Road to Victory: Coalition Formation between the Suffragists and the Woman's Christian Temperance Union. *Mobilization* 7:231–251.

McClintock, Nathan. 2011. From Industrial Garden to Food Desert: Demarcated Devalution in the Flatlands of Oakland, California. In *Cultivating Food Justice:*

Race, Class, and Sustainability, ed. A. Alkon and J. Agyeman. Cambridge, Mass.: MIT Press.

McClure, Kirk. 2008. Deconcentrating Poverty with Housing Programs. *Journal of the American Planning Association* 74 (1):90–99.

McFarlane, Colin. 2010. The Comparative City: Knowledge, Learning, Urbanism. *International Journal of Urban and Regional Research* 34 (4):725–742.

McGurty, Eileen. 2000. Warren County, NC, and the Emergence of the Environmental Justice Movement: Unlikely Coalitions and Shared Meanings in Local Collective Action. *Society & Natural Resources* 13 (4):373–387.

McGurty, Eileen Maura. 2007. *Transforming Environmentalism: Warren County, PCBS, and the Origins of EnvironmentalJjustice*. New Brunswick, N.J.: Rutgers University Press.

Medoff, Peter, and Holly Sklar. 1994. *Streets of Hope: The Fall and Rise of an Urban Neighborhood*. Boston: South End Press.

Melo, Carme. 2006. Towards an Urban Ecological Citizenship. *Re-public*. http://www.re-public.gr/en/?p=43.

Merrifield, Andy, and Erik Swyngedouw. 1997. *The Urbanization of Injustice*. New York: New York University Press.

Mitchell, Don. 2003. *The Right to the City: Social Justice and the Fight for Public Space*. New York: Guilford Press.

Mitchell, Gordon, and Danny Dorling. 2003. An Environmental Justice Analysis of British Air Quality. *Environment & Planning A* 35 (5):909–929.

Molan, Gabriella. 2007. *Turf Wars: Discourse, Diversity, and the Politics of Place*. Malden, Mass.: Blackwell.

Mollenkopf, John H. 1981. Community and Accumulation. In *Urbanization and Urban Planning in Capitalist Society*, ed. M. J. Dear and A. J. Scott. New York: Methuen.

Monnet, Nadja. 2002. *La formación del espacio público: Una mirada etnológica sobre el casc antic de Barcelona*. Barcelona: Los Libros de la Catarata.

Montagna, Nicola. 2006. The De-commodification of Urban Space and the Occupied Social Centres in Italy 1. *City* 10 (3):295–304.

Montaner, Jose Maria. 2004. La evolución del modelo Barcelona (1979–2002). In *Urbanismo en el siglo XXI: Una visión crítica: Bilbao, Madrid, Valencia, Barcelona*, ed. J. Borja, Z. Muxí and J. Cenicacelaya. Barcelona: Escola Tècnica Superior d'Arquitectura de Barcelona, Edicions UPC.

Moore, Latetia, and Ana Diez Roux. 2006. Associations of Neighborhood Characteristics with the Location and Type of Food Stores. *American Journal of Public Health* 96 (2):325–331.

Morales Dominguez, Esteban. 2004. Cuba: Los retos del color. In *Heterogeneidad social en la Cuba actual*, ed. P. Villanueva, L. Iñiquez Rojas, and O. Everterry. Havana: Universidad de La Habana, Centro de Estudios de la Salud y Bien Estar Humano.

Morello-Frosch, Rachel. 2002. Discrimination and the Political Economy of Environmental Inequality. *Environment and Planning. C, Government & Policy* 20 (4):477–496.

Morland, Kimberly, Steve Wing, and Ana Diez Roux. 2002. The Contextual Effect of the Local Food Environment on Residents' Diets: The Atherosclerosis Risk in Communities Study. *American Journal of Public Health* 92 (11):1761–1768.

Morland, Kimberly, Steve Wing, Ana Diez Roux, and Charles Poole. 2002. Neighborhood Characteristics Associated with the Location of Food Stores and Food Service Places. *American Journal of Preventive Medicine* 22 (1):23–29.

Nello, Oriol. 2004. Urban Dynamics, Public Policies, and Governance in the Metropolitan Region of Barcelona. In *Transforming Barcelona*, ed. T. Marshall. London: Routledge.

Newell, Peter. 2005. Race, Class and the Global Politics of Environmental Inequality. *Global Environmental Politics* 5 (3):70–94.

O'Connor, Alice. 2001. Understanding Inequality in the Late Twentieth-Century Metropolis: New Perspectives on the Enduring Racial Divide. In *Urban Inequality in the United States: Evidence from Four Cities*, ed. L. Bobo, A. O'Connor, and C. Tilly. New York: Russell Sage Foundation.

Oliveras, Rosa, and Regla Díaz. 2007. Hacer ciudad y hacer barrios: Los Talleres de Transformación Integral del Barrio. In *Cultura, tradición, y comunidad: Perspectivas sobre la participación y el desarrollo en Cuba*, ed. A. H. Hearn. Havana: Imagen Contemporánea.

Ortega Cerdà, Miquel, and Maria Calaf Forn. 2010. Equitat ambiental a Cataluya: Integració de les dimensions ambiental, territorial, i social a la presa de decisions. In *Documents de recerca*, ed. Consell Assessor per al Desenvolupament Sostenible y Generalitat de Catalunya. Barcelona: Generalitat de Catalunya.

Ortega, Elisenda. 2008. *Proposta per a l'espai comunitari Pou de la Figuera*. Barcelona: Ajuntament de Barcelona.

Otero, Lydia. 2010. *La Calle: Spatial Conflicts and Urban Renewal in a Southwest City*. Tucson: University of Arizona Press.

Pachucki, Mark, Sabrina Pendergrass, and Michelle Lamont. 2007. Boundary Processes: Recent Theoretical Developments and New Contributions. *Poetics* 35 (6):331–351.

Park, Lisa Sun-Hee, and David Pellow. 2011. *The Slums of Aspen: Immigrants vs. the Environment in America's Eden*. New York: New York University Press.

Parks, Bradley, and J. Timmons Roberts. 2006. Globalization, Vulnerability to Climate Change, and Perceived Injustice in the South. *Society & Natural Resources* 19 (4):337–355.

Pattillo, Mary. 2007. *Black on the Block: The Politics of Race and Class in the City*. Chicago: University of Chicago Press.

Paz, A. 2004. *La Barcelona Rebelde. Guía de una Ciudad Silenciada*. Barcelona: Octaedro.

Pearsall, Hamil. 2012. Moving Out or Moving In? Resilience to Environmental Gentrification in New York City. *Local Environment* 17 (9):1013–1026.

Pellow, David. 2000. Environmental Inequality Formation: Toward a Theory of Environmental Justice. *American Behavioral Scientist* 43 (4):581–601.

Pellow, David. 2001. Environmental Justice and the Political Process: Movements, Corporations, and the State. *Sociological Quarterly* 42 (1):47–67.

Pellow, David. 2002. *Garbage Wars: The Struggle for Environmental Justice in Chicago.* Cambridge, Mass.: MIT Press.

Pellow, David. 2007. *Resisting Global Toxics: Transnational Movements for Environmental Justice.* Cambridge, Mass.: MIT Press.

Pellow, David. 2009. "We Didn't Get the First Five Hundred Years Right, so Let's Work on the Next Five Hundred Years": A Call for Transformative Analysis and Action. *Environmental Justice* 2 (1):3–6.

Pellow, David, and Robert J. Brulle. 2005. *Power, Justice, and the Environment: A Critical Appraisal of the Environmental Justice Movement.* Cambridge, Mass.: MIT.

Pellow, David N., and Lisa Sun-Hee Park. 2002. *The Silicon Valley of Dreams: Environmental Injustice, Immigrant Workers, and the High-Tech Global Economy.* New York: New York University Press.

Pérez-López, Jorge, and Sergio Díaz-Briquets. 2000. *Conquering Nature: the Environmental Legacy of Socialism in Cuba.* Pittsburgh: Pittsburgh University Press.

Perkins, Douglas, and Adam Long. 2002. Neighborhood Sense of Community and Social Capital: A Multi-level Analysis. In *Psychological Sense of Community: Research, Applications, and Implications*, ed. A. Fisher, C. Sonn, and B. Bishop. New York: Plenum.

Peterman, William. 2000. *Neighborhood Planning and Community-based Development: The Potential and Limits of Grassroots Action.* Thousand Oaks, Calif.: Sage.

Polletta, Francesca. 1999. "Free Spaces" in Collective Action. *Theory and Society* 28 (1):1–38.

Polletta, Francesca. 2005. How Participatory Democracy Became White: Culture and Organizational Choice. *Mobilization* 10 (2):271–288.

Porta, D. d., and D. Rucht. 2002. The Dynamics of Environmental Campaigns. *Mobilization* 7 (10):1–14.

Project Right. 2008. *Reflections from the Past . . . Organizing for the Future: Annual Report 2007–2009.* Dorchester: Project Right.

Proshansky, Harold, Abbe Fabian, and Robert Kaminoff. 1983. Place Identity: Physical World Socialization of the Self. *Journal of Environmental Psychology* 3 (1):57–83.

Pulido, Laura. 1996. *Environmentalism and Economic Justice: Two Chicano Struggles in the Southwest.* Tucson: University of Arizona Press.

Purcell, Mark. 2008. *Recapturing Democracy: Neoliberalization and the Struggle for Alternative Urban Futures.* New York: Routledge.

Putnam, Robert. 2001. Let's Play Together. *The Observer* 25.

Putnam, Robert, Lewis Feldstein, and Don Cohen. 2004. *Better Together: Restoring the American Community.* New York: Simon & Schuster.

Ramirez, R. 2005. State and Civil Society in the Barrios of Havana, Cuba: The Case of Pogolotti. *Environment and Urbanization* 17 (1):147–170.

Rey, Gina. 2001. *La Habana: Del barrio a la ciudad.* Washington, D.C.: Latin American Studies Association.

Rey, Gina, Jorge Peña, and Tania Gutiérrez. 2006. *Proyecto de investigacion: Rehabilitacion urbana sustentable para Centro Habana. Parte 1: Marco Conceptual.* Havana: Centro de Estudios Urbanos (CEU-H), Facultad de Arquitectura.

Riger, Stephanie, and Paul Lavrakas. 1981. Community Ties: Patterns of Attachment and Social Interaction in Urban Neighborhoods. *American Journal of Community Psychology* 9 (1):55–66.

Robbins, Paul. 2011. *Political Ecology: A Critical Introduction.* Chichester, UK: Wiley-Blackwell.

Robinson, Jennifer. 2011. Cities in a World of Cities: The Comparative Gesture. *International Journal of Urban and Regional Research* 35 (1):1–23.

Rodrigues, M. G. M. 2004. Advocating for the Environment—Local Dimensions of Transnational Networks. *Environment* 46 (2):14–25.

Rosenthal, Robert. 1976. *Different Strokes: Pathways to Maturity in the Boston Ghetto. A Report to the Ford Foundation.* Boulder, Colo.: Westview Press.

Roy, Ananya. 2011. Slumdog Cities: Rethinking Subaltern Urbanism. *International Journal of Urban and Regional Research* 35 (2):223–238.

Saegert, Susan, J. Philip Thompson, and Mark Warren. 2001. *Social Capital and Poor Communities.* New York: Russell Sage Foundation.

Sampson, Robert. 2004. Neighborhood and Community: Collective Efficacy and Community Safety. *New Economy* 11:106–113.

Sánchez Lopez, Pere. 1986. *El centro histórico: Un lugar para el conflicto. Estrategias del capital para la expulsión del proletario del centro de Barcelona. El caso de Santa Caterina y el Portal Nou.* Barcelona: Publicacions i Edicions de la Universitat de Barcelona.

Sandercock, Leonie. 1998. *Towards Cosmopolis: Planning for Multicultural Cities.* Chichester, UK: Wiley.

Sandercock, Leonie. 2003. Toward Cosmopolis: Utopia as Construction Site. In *Readings in Planning Theory,* ed. S. Campbell and S. Fainstein. Malden, Mass.: Wiley-Blackwell.

Sandler, Ronald D., and Phaedra C. Pezzullo. 2007. *Environmental Justice and Environmentalism: The Social Justice Challenge to the Environmental Movement.* Cambridge, Mass.: MIT Press.

San Sebastián, Miguel, B. Armstrong, J. A. Córdoba, and C. Stephens. 2001. Exposures and Cancer Incidence Near Oil Fields in the Amazon Basin of Ecuador. *Occupational and Environmental Medicine* 58 (8):517–522.

Sardiñas, Rosa. 2009. Un pequeño lugar en el mundo. Casa del Niño y de la Niña, Havana.

Sassen, Saskia. 1998. Globalization and Its Discontents. In *The Blackwell City Reader*, ed. G. Bridge and S. Watson. New York: Blackwell.

Sassen, Saskia. 1999. Whose City Is It? Globalization and the Formation of New Claims. In *The Urban Moment: Cosmopolitan Essays on the Late Twentieth-Century City*, ed. R. A. Beauregard and S. Body-Gendrot. Thousand Oaks, Calif.: Sage.

Satterthwaite, D., S. Huq, H. Reid, M. Pelling, and P. Romero-Lankao. 2007. *Adapting to Climate Change in Urban Areas: The Possibilities and Constraints in Low- and Middle-Income Nations*. London: International Institute for Environment and Development.

Scannell, Leila, and Robert Gifford. 2010. The Relations between Natural and Civic Place Attachment and Pro-environmental Behavior. *Journal of Environmental Psychology* 30 (3):289–297.

Scarpaci, Joseph L., Roberto Segre, and Mario Coyula. 2002. *Havana: Two Faces of the Antillean Metropolis*. Rev. ed. Chapel Hill: University of North Carolina Press.

Schlosberg, David. 2007. *Defining Environmental Justice: Theories, Movements, and Nature*. Oxford: Oxford University Press.

Schmelzkopf, Karen. 2002. Incommensurability, Land Use, and the Right to Space: Community Gardens in New York City. *Urban Geography* 23 (4):323–343.

Schnaiberg, Allan. 1980. *The Environment: From Surplus to Scarcity*. New York: Oxford University Press.

Schnaiberg, Allan, and K. Gould. 1994. *Environment and Society: The Enduring Conflict*. New York: St. Martin's Press.

Schnaiberg, Allan, David Pellow, and Adam Weinberg. 2002. The Treadmill of Production and the Environmental State. In *Research in Social Problems and Public Policy*, ed. A. Mol and F. Buttel. Greenwich, Conn.: Emerald.

Schneider, Stephen. 2007. *Refocusing Crime Prevention: Collective Action and the Quest for Community*. Toronto: University of Toronto Press.

Self, Robert O. 2003. *American Babylon: Race and the Struggle for Postwar Oakland: Politics and Society in Twentieth-Century America*. Princeton, N.J.: Princeton University Press.

Sennett, Richard. 2007. A Flexible City of Strangers. *ARC* 66:19–23.

Settles, Patricia. 1994. *Community Involvement in Hazardous Waste Site Cleanup in the Dudley Street Neighborhood*. Medford, Mass.: Tufts University.

Shaw, K. 2004. The Global/local Politics of the Great Bear rainforest. *Environmental Politics* 13 (2):371–392.

Shiva, Vandana. 2002. *Water Wars: Privatization, Pollution and Profit.* Cambridge, Mass.: South End Press.

Shiva, Vandana, and J. Bandyopadhyay. 1991. *Ecology and the Politics of Survival: Conflicts over Natural Resources in India.* New Delhi: Sage.

Shiva, Vandana, and Gitanjali Bedi. 2002. *Sustainable Agriculture and food Security: The Impact of Globalisation.* New Delhi: Sage.

Shrader-Frechette, K. S. 2007. *Taking Action, Saving Lives: Our Duties to Protect Environmental and Public Health.* Oxford: Oxford University Press.

Shutkin, William. 2000. *The Land That Could Be: Environmentalism and Democracy in the Twenty-first Century.* Cambridge, Mass.: MIT Press.

Sinclair, Minor, and Martha Thompson. 2001. *Cuba, Going against the Grain: Agricultural Crisis and Transformation.* Boston: Oxfam America.

Small, Mario Luis. 2004. *Villa Victoria: The Transformation of Social Capital in a Boston Barrio.* Chicago: University of Chicago Press.

Smith, Neil. 1982. Gentrification and Uneven Sevelopment. *Economic Geography* 58:139–155.

Smith, Neil. 1986. Gentrification, the Frontier and the Restructuring of urban Apace. In *Gentrification in the City,* ed. N. Smith and P. Williams. London: Unwin Hyman.

Smith, Neil. 1987. Gentrification and the Rent Gap. *Annals of the Association of American Geographers* 77 (3):462–465.

Smith, Ted, David Allan Sonnenfeld, and David N. Pellow. 2006. *Challenging the Chip: Labor Rights and EnvironmentalJjustice in the Global Electronics Industry.* Philadelphia: Temple University Press.

Smoyer-Tomic, Karen, John Spence, and Carl Amrhein. 2006. Food Deserts in the Prairies? Supermarket Accessibility and Neighborhood Need in Edmonton, Canada. *Professional Geographer* 58 (3):307–326.

Soja, Edward. 2000. *Postmetropolis: Critical Studies of Cities and Regions.* Oxford: Blackwell.

Soja, Edward. 2009. The City and Spatial Justice. *Spatial Justice* 1:31–38.

Soja, Edward. 2010. *Seeking Spatial Justice.* Minneapolis: University of Minnesota Press.

Spiegel, Jerry, Mariano Bonet, Maricel Garcia, Ana Maria Ibarra, Robert Tate, and Annalee Yassi. 2004. Building Capacity in Central Havana to Sustainably Manage Environmental Health Risk in an Urban Ecosystem. *EcoHealth* 1 (2):120–130.

Spiegel, Jerry, Mariano Bonet, Annalee Yassi, Enrique Molina, Miriam Concepcion, and Pedro Mast. 2001. Developing Ecosystem Health Indicators in Centro Habana: A Community-Based Approach. *Ecosystem Health* 7 (1):15–26.

Squires, Gregory. 1991. Partnership and the Pursuit of the Private Vity. In *Urban Life in Transition,* ed. M. Gottdiener and C. G. Pickvance. London: Sage.

Staggenborg, Suzanne. 1986. Coalition Work in the Pro-Choice Movement: Organizational and Environmental Opportunities and Obstacles. *Social Problems* 33:374–390.

Staggenborg, Suzanne. 1998. The Consequences of Professionalization and Formalization in the Pro-choice Movement. *American Sociological Review* 53:585–606.

Steil, Justin, and James Connolly. 2009. Can the Just City Be Built from Below? Brownfields, Planning, and Power in the South Bronx. In *Searching for the Just City: Debates in Urban Theory and Practice*, ed. Peter Marcuse, James Connolly, Johannes Novy, Ingrid Olivo, Cuz Potter, and Justin Steil. London: Routledge.

Stein, Rachel. 2004. *New Perspectives on Environmental Justice: Gender, Sexuality, and Activism*. New Brunswick, N.J.: Rutgers University Press.

Sugrue, Thomas. 2005. *The Origins of the Urban Crisis: Race and Inequality in Postwar Detroit*. Princeton, N.J.: Princeton University Press.

Suttles, Gerald. 1968. *The Social Order of the Slum: Ethnicity and Territory in the Inner City*. Chicago: University of Chicago Press.

Sze, J. 2007. *Noxious New York: The Racial Politics of Urban Health and Environmental Justice*. Cambridge, Mass.: MIT Press.

Takano, Guillermo, and Juan Tokeshi. 2007. Espacio público en la ciudad popular: Reflexiones y experiencias desde el Sur. *Estudios Urbanos*. Lima: DESCO.

Taller contra la Violencia Inmobiliaria y Urbanística. 2006. *El cielo está enladrillado: Entre el "mobbing" y la violencia inmobiliaria y urbanística*. Barcelona: Ediciones Bellaterra.

Tarrow, Sidney. 1994. *Power in Movement: Social Movements, Collective Action, and Politics*. Cambridge, U.K.: Cambridge University Press.

Tarrow, Sidney. 2011. *Power in Movement: Social Movements and Contentious Politics*. 2nd ed. New York: Cambridge University Press.

Taylor, Henry. 2009. *Inside El Barrio: A Bottom-Up View of Neighborhood Life in Castro's Cuba*. Sterling, Va.: Kumarian Press.

Tello, Enric. 2004. Changing Course? Principles and Tools for Local Sustainability. In *Transforming Barcelona*, ed. T. Marshall. London: Routledge.

Tills, Karen. 2005. *The New Berlin: Memory, Politics, Place*. Minneapolis: University of Minnesota Press.

Tilly, Charles. 1974. Do Communities Act? In *The Community: Approaches and Applications*, ed. M. P. Effrat. New York: Free Press.

Tilly, Charles. 1978. *From Mobilization to Revolution*. Reading, Mass.: Addison-Wesley.

Tilly, Charles. 2001. *Coercion, Capital, and European States, AD 990–1992*. Cambridge, MA: Blackwell.

Tilly, Charles. 2006. *Regimes and Repertoires*. Chicago: University of Chicago Press.

Tönnies, Ferdinand. 1957. *Community and Society (Gemeinschaft und Gesellschaft)*. East Lansing: Michigan State University Press.

Turner, Margery. 1998. Moving out of Poverty: Expanding Mobility and Choice through Tenant-Based Housing Assistance. *Housing Policy Debate* 9 (2):373–394.

Unió Temporal d'Escribes. 2004. *Barcelona, Marca Registrada: Un Model per Desarmar*. Barcelona: Virus.

Uriarte-Gastón, Miren. 2002. *Cuba, Social Policy at the Crossroads: Maintaining Priorities, Transforming Practice*. Boston: Oxfam America.

Van Dyke, Nella. 2003. Crossing Movement Boundaries: Factors That Facilitate Coalition Protest by American College Students, 1930–1990. *Social Problems* 50 (2):226–250.

Varga, Csaba, Istvan Kiss, and Istvan Ember. 2002. The Lack of Environmental Justice in Central and Eastern Europe. *Environmental Health Perspectives* 110 (11):662–663.

Venkatesh, Sudhir. 2000. *American Project: The Rise and Fall of a Modern Ghetto*. Cambridge, Mass.: Harvard University Press.

Ver Ploeg, Michele, Vince Breneman, Tracey Farrigan, Karen Hamrick, David Hopkins, Phil Kaufman, Biing-Hwan Lin, et al. 2009. *Access to Affordable and Nutritious Food: Measuring and Understanding Food Deserts and Their Consequences*. Washington, D.C.: U.S. Department of Agriculture.

Vogel, David. 2006. The Private Regulation of Global Corporate Conduct. Paper presented at Annual Meeting of the American Political Science Association, Philadelphia, September 3.

Von Hoffman, Alexander. 2003. *House by House, Block by Block: The Rebirth of America's Urban Neighborhoods*. New York: Oxford University Press.

Wacquant, Loïc. 2007. Territorial Stigmatization in the Age of Advanced Marginality. *Thesis Eleven* 91 (1): 66–77.

Warner, Sam Bass. 1978. *Streetcar Suburbs: The Process of Growth in Boston, 1870–1900*. 2nd ed. Cambridge, Mass.: Harvard University Press.

Warner, Sam Bass, and Hansi Durlach. 1987. *To Dwell Is to Garden: A History of Boston's Community Gardens*. Boston: Northeastern University Press.

Warren, Mark. 2001. *Dry Bones Rattling: Community Building to Revitalize American Democracy*. Princeton, N.J.: Princeton University Press.

Weinstein, Liza. 2008. Mumbai's Development Mafias: Globalization, Organized Crime and Land Development. *International Journal of Urban and Regional Research* 32 (1):22–39.

Wheatley, Brian, and Margaret Wheatley. 2000. Methylmercury and the Health of Indigenous Peoples: A Risk Management Challenge for Physical and Social Sciences and for Public Health Policy. *Science of the Total Environment* 259 (1–3):23–29.

Wheeler, Benedict W, and Yoav Ben-Shlomo. 2005. Environmental Equity, Air Quality, Socioeconomic Status, and Respiratory Health: A Linkage Analysis of Routine Data from the Health Survey for England. *Journal of Epidemiology and Community Health* 59 (11):948–954.

Wrigley, Neil, Daniel Warm, Barrie Margetts, and Michelle Lowe. 2004. The Leeds "Food Deserts" Intervention Study: What the Focus Groups Reveal. *International Journal of Retail & Distribution Management* 32 (2–3):123.

Wu, Fulong. 2000. The Global and Local Dimensions of Place Making: Remaking Shanghai as a World City. *Urban Studies* 37 (8):1359–1377.

Wu, Fulong. 2004. Urban poverty and marginalization under market transition: the case of Chinese cities. *International Journal of Urban and Regional Research* 28 (2):401–423.

Yassi, Annalee, Niurys Fernandez, Ariadna Fernandez, Mariano Bonet, Robert Tate, and Jerry Spiegel. 2003. Community Participation in a Multisectoral Intervention to Address Health Determinants in an Inner-City Community in Central Havana. *Journal of Urban Health Research* 80 (1):61–80.

Yassi, Annalee, Pedro Mas, Mariano Bonet, Robert Tate, Niurys Fernandez, Jerry Spiegel, and Mayilee Perez. 1999. Applying an Ecosystem Approach to the Determinants of Health in Centro Habana. *Ecosystem Health* 5 (1):3–19.

Yiftachel, Oren. 2009. Theoretical Notes on "Gray Cities": The Coming of Urban Apartheid? *Planning Theory* 8 (1):88–100.

Yin, R. K. 2003. *Case Study Research: Design and Methods.* 3rd ed. Thousand Oaks, Calif.: Sage Publications.

Young, Iris Marion. 1990. *Justice and the Politics of Difference.* Princeton, N.J.: Princeton University Press.

Zald, Mayer, and John McCarthy. 1987. *Social Movements in an Organizational Society.* New Brunswick, N.J.: Transaction Publishers.

Index

Urban and Industrial Environments

Series editor: Robert Gottlieb, Henry R. Luce Professor of Urban and Environmental Policy, Occidental College

Steve Lerner, *Diamond: A Struggle for Environmental Justice in Louisiana's Chemical Corridor*

Jason Corburn, *Street Science: Community Knowledge and Environmental Health Justice*

Peggy F. Barlett, ed., *Urban Place: Reconnecting with the Natural World*

David Naguib Pellow and Robert J. Brulle, eds., *Power, Justice, and the Environment: A Critical Appraisal of the Environmental Justice Movement*

Eran Ben-Joseph, *The Code of the City: Standards and the Hidden Language of Place Making*

Nancy J. Myers and Carolyn Raffensperger, eds., *Precautionary Tools for Reshaping Environmental Policy*

Kelly Sims Gallagher, *China Shifts Gears: Automakers, Oil, Pollution, and Development*

Kerry H. Whiteside, *Precautionary Politics: Principle and Practice in Confronting Environmental Risk*

Ronald Sandler and Phaedra C. Pezzullo, eds., *Environmental Justice and Environmentalism: The Social Justice Challenge to the Environmental Movement*

Julie Sze, *Noxious New York: The Racial Politics of Urban Health and Environmental Justice*

Robert D. Bullard, ed., *Growing Smarter: Achieving Livable Communities, Environmental Justice, and Regional Equity*

Ann Rappaport and Sarah Hammond Creighton, *Degrees That Matter: Climate Change and the University*

Michael Egan, *Barry Commoner and the Science of Survival: The Remaking of American Environmentalism*

David J. Hess, *Alternative Pathways in Science and Industry: Activism, Innovation, and the Environment in an Era of Globalization*

Peter F. Cannavò, *The Working Landscape: Founding, Preservation, and the Politics of Place*

Paul Stanton Kibel, ed., *Rivertown: Rethinking Urban Rivers*

Kevin P. Gallagher and Lyuba Zarsky, *The Enclave Economy: Foreign Investment and Sustainable Development in Mexico's Silicon Valley*

David N. Pellow, *Resisting Global Toxics: Transnational Movements for Environmental Justice*

Robert Gottlieb, *Reinventing Los Angeles: Nature and Community in the Global City*

David V. Carruthers, ed., *Environmental Justice in Latin America: Problems, Promise, and Practice*

Tom Angotti, *New York for Sale: Community Planning Confronts Global Real Estate*

Paloma Pavel, ed., *Breakthrough Communities: Sustainability and Justice in the Next American Metropolis*

Anastasia Loukaitou-Sideris and Renia Ehrenfeucht, *Sidewalks: Conflict and Negotiation over Public Space*

David J. Hess, *Localist Movements in a Global Economy: Sustainability, Justice, and Urban Development in the United States*

Julian Agyeman and Yelena Ogneva-Himmelberger, eds., *Environmental Justice and Sustainability in the Former Soviet Union*

Jason Corburn, *Toward the Healthy City: People, Places, and the Politics of Urban Planning*

JoAnn Carmin and Julian Agyeman, eds., *Environmental Inequalities beyond Borders: Local Perspectives on Global Injustices*

Louise Mozingo, *Pastoral Capitalism: A History of Suburban Corporate Landscapes*

Gwen Ottinger and Benjamin Cohen, eds., *Technoscience and Environmental Justice: Expert Cultures in a Grassroots Movement*

Samantha MacBride, *Recycling Reconsidered: The Present Failure and Future Promise of Environmental Action in the United States*

Andrew Karvonen, *Politics of Urban Runoff: Nature, Technology, and the Sustainable City*

Daniel Schneider, *Hybrid Nature: Sewage Treatment and the Contradictions of the Industrial Ecosystem*

Catherine Tumber, *Small, Gritty, and Green: The Promise of America's Smaller Industrial Cities in a Low-Carbon World*

Sam Bass Warner and Andrew H. Whittemore, *American Urban Form: A Representative History*

John Pucher and Ralph Buehler, eds., *City Cycling*

Stephanie Foote and Elizabeth Mazzolini, eds., *Histories of the Dustheap: Waste, Material Cultures, Social Justice*

David J. Hess, *Good Green Jobs in a Global Economy: Making and Keeping New Industries in the United States*

Joseph F. C. DiMento and Clifford Ellis, *Changing Lanes: Visions and Histories of Urban Freeways*

Joanna Robinson, *Contested Water: The Struggle against Water Privatization in the United States and Canada*

William B. Meyer, *The Environmental Advantages of Cities: Countering Commonsense Antiurbanism*

Rebecca L. Henn and Andrew J. Hoffman, eds., *Constructing Green: The Social Structures of Sustainability*

Peggy F. Barlett and Geoffrey W. Chase, eds., *Sustainability in Higher Education: Stories and Strategies for Transformation*

Isabelle Anguelovski, *Neighborhood as Refuge: Community Reconstruction, Place Remaking, and Environmental Justice in the City*